Handbook For The Deep Ecologist

Handbook For The Deep Ecologist

✦

What Everyone Should Know About Self, the Environment, and the Planet

Dr. Scott J. Neuhaus

iUniverse, Inc.
New York Lincoln Shanghai

Handbook For The Deep Ecologist
What Everyone Should Know About Self, the Environment, and the Planet

iUniverse books may be ordered through booksellers or by contacting:

iUniverse
2021 Pine Lake Road, Suite 100
Lincoln, NE 68512
www.iuniverse.com
1-800-Authors (1-800-288-4677)

ISBN-13: 978-0-595-35789-5 (pbk)
ISBN-13: 978-0-595-80258-6 (ebk)
ISBN-10: 0-595-35789-X (pbk)
ISBN-10: 0-595-80258-3 (ebk)

Printed in the United States of America

Contents

Introduction

I was struck by an article in the newspaper the other day that was talking about the new hybrid technology. It concerned a large American automobile manufacturing company and its approach to implementing and marketing the innovation. Their strategy was not to demonstrate an advanced ability to conserve gas and release less emissions thereby lessening our total demand nationally and stress on the planet generally. Instead, the entire marketing ploy was that the technology, in large SUV's, was yielding an amazing 20+ miles per gallon. So now the approach of the commercials was "see, you can have a large gas guzzling SUV, but without the guilt of owning a large, gas guzzling SUV". This national love affair that we as Americans have with our cars and trucks is reaching beyond embarrassing. It's beyond an obsession. It has become a sign of American economic arrogance around the world and we, apparently, really don't care.

The purpose of this book is hopefully to make Americans care more. By first understanding ourselves and how we came to be, not Americans, but simply, just members of the world's society of plants and animals, only then can it start.

By mentioning the problem as it is with American cars and their owners, I certainly don't mean to say that the rest of the world's inhabitants have nothing to learn, though. The concepts inherent in the philosophy of Deep Ecology (I'm not a Deep Ecologist, think they are a bit extreme, but I get their message) can only be implemented if we understand who we as organisms are, why we react to certain stimuli the way we do, why certain concepts are difficult to grasp, why certain things are more important than others, how over time we came to be this way, how our planet came to be the way it is, and how our civilizations came to be the way they are.

Through understanding of these aforementioned points and then by integrating the cycles of life regarding the essential elements to life on earth, we can begin to consider the planet as what it really is…an island in the boundlessness of space, with finite resources, a finite ability to withstand environmental stress, that is home to many, many species of plant and animal, of which we are just one.

I asked my wife Joan the other day, in passing, that if it were possible to reason and communicate as humans do, what would nearly every species of animal and a preponderance of plant species include in a "List of things to do in order to survive on earth"? She just thought briefly and replied stoically, "Stay away from humans!" I had considered the same question and came up with the same thought only moments before. Think about it though, and it will seem rather obvious. Now, granted, in the history of our planet's existence, probably 90% or more of all species that ever existed, never had the occasion to co-exist with *Homo sapiens*. So obviously species come and species go as is the natural way of the evolutionary process. Nature doesn't get teary eyed when a species fails to perform "properly" and is lost forever.

Also, there have even been periods of mass extinction due to cataclysmic events several times over the last 500 million years or so. Cataclysmic events causing extinctions is one thing, though. Extinctions of many species caused merely by direct human predation or side effects of human overpopulation, is quite another.

This all stems from our "great leap forward" where we acquired the ability to reason some 40,000 to 80,000 years ago. The advantage this quirk of the evolutionary process yielded, had a side effect. Since we could think and no other species of plant or animal could think and reason like we could, then everything had to have been put here for us, right? It must be so, that we *are* the chosen species! *Anthropocentrism* (all life revolves around man's existence) was born. This was and continues to be the largest obstacle in the path of education of the world's inhabitants, as to the threat we present to the world and to ourselves.

The way this book is organized is itself a bit anthropocentric in that it covers initially the evolution of *Homo sapiens* and the "tools" that we have at our disposal today due to the process of natural selection through millions of years. It discusses these outdated strategies that served us so well during our hunter-gatherer period, and how they are played to and engaged by marketing approaches that try to separate you from your money. Our standardized sleep approach is discussed showing that we are *directed* more than we *direct* when it comes to most biological entities in our lives. Other topics follow along the same vein. I believe it is crucial to understand how "tied to the earth" we are as individuals of a species, before we can understand our place as merely part of the fabric of life on this planet.

The cycles of life as they relate to the six most essential elements of the living processes are discussed. Great care was taken to overwhelm the reader as to just how important, not only as components of these cycles but also in other aspects of our lives, each elements is. True appreciation of their importance can be comprehended especially by measuring the imbalance humans have caused in their cycles.

The roles of bacteria and viruses are included due to the part they play in the symptomology of our troubled ecosphere. Due to ever increasing human populations and the resulting putrefaction of parts of our environment, ever increasing populations of microbes stand poised to display their "naturally corrective effects" on our numbers. By understanding the nuances of the human-microbe-environment relationship we can better control our vulnerability to them.

The Geology section was included to highlight shortcomings in both the way we view our planet and the way we conceptualize time. The fact that the island we inhabit in the void of space is both dynamic and long-lived should make one consider the human-centered approach we have to our "home". The science that establishes timeframes of geology is similar to the science in other aspects of our lives we take for granted. It's not merely a contrived conspiracy to refute the existence of God (for it doesn't do that anyway, unless you want it to). The same can be said for the other "forensic" tools used to study the past and the conclusions they suggest.

The civilization portion became important to provide a timeframe for our existence on the planet, lessons in failure, and lessons in success when it comes to harmonizing and succeeding. Included were archeological assertions as to when different plants and animals were domesticated, why this occurred, and their importance to our species through time. This section is particularly foreboding when considering that certain civilizations, like past species, failed to adapt, thereby becoming extinct themselves, as a society or civilization.

The environmental impact of plant and animal species introduction into non-native environs, either as part of the domestication process or merely as secondary effects of improved global transportation, is examined in the Alien section. The myriad of problems associated with the examples discussed as well as those omitted underscores the folly of human arrogance. The SETI portion, and the evacuation possibilities of our species from this planet, was added to underscore the fact that we can't really go anywhere, so it would behoove us to manage our ecosphere better.

Finally a discussion on the interrelationship of it all in the context of Deep Ecology is presented. Assessments and predictions are made but nothing really is accomplished except, hopefully, what I set out to do which is simply to make clearer where we came from as a species *and* as a planet, where we are as a species *and* as a planet, and where we can expect to end up as a species *and* as a planet, if the right steps are taken versus if we ignore the symptoms of a sick planet.

So am I a world-recognized expert in the field of Ecology, Chemistry, or History? No, I am not. My publishing company editor, after analyzing my work, indicated to me that the book was very interesting and easy to read with important material but that I needed to say why people would want to listen to what I had to say. Specifically I needed to "convince readers of" my "expertise in this area" and present my "credentials in a way that will carry more weight with readers". As much as I love and respect the people at *iUniverse* I disagreed with this because this is information that is already out there in abundance. My baccalaureate degree in biochemistry and graduate education I believe only gives me the tools to understand the complex science that's out there along with the bounty of not-so-complex stuff. Fourteen months of research and compilation doesn't make me an expert anymore than my education. But I did have fun writing it. It's not even close to the specific exhaustive works of John McPhee, Laurie Garret, or Jared Diamond (who are cited in my bibliography and are my heroes, by the way). But more often than not people want a work like mine. A Handbook that can be read quickly that can allow an epiphany. I believe it's unique as a work because it will allow a superstructure for future knowledge the reader is exposed to "across the board", and it truly is a formulation of work that, I believe everyone should know to be a 21st century human.

Acknowledgements

First and foremost I want to thank my wife, Joan, who is my love and companion during my stay here on this planet. Her warm-heartedness and goodness are an inspiration to me.

I couldn't have done this without my stepson, Luke Ware, who did all things technical for me, especially as it regards my website, from which all things grew.

I also want to express gratitude to my oldest brother, Barry, whose positive influence at the right time of my life made such a difference.

Last, I can't forget my mother, who instilled in me an attitude of not being afraid to try things, and a work ethic to propel it.

Prologue

The family *Hominidae* (*Hominids*), of which *Homo sapiens* (humans) are members, is of course mammalian. The evolution of mammals, after the critical main branching from their progenitors, really got started at the end of the Permian era (253 million years ago) due to what happened during what is called the *end Permian extinction*. During this time 90% of all fauna and 60% of all Genera went extinct! This was thought to occur because of a radical change in the environment due to either great volcanic activity, collision of the earth with a large meteor, the appearance of a pangea (one very large continent) with loss of continental shelf habitat or a combination of the three.

After this, mammalian fossils began to appear in the geologic strata. They were small and shrew-like and corresponded with the appearance of dinosaurs. Unfortunately due to the success of the dinosaurs, mammals occupied a much smaller niche than they would come to acquire later and this was due directly to the predation affects of dinosaurs. Dinosaurs liked to eat mammals. As a result, mammalian size and range was kept in check. Apparently larger mammals were just too vulnerable. Adapting mammals were "selected for" to be mostly nocturnal.

Things changed dramatically 65 million years ago (*Cretaceous extinction*). This is when an impact of the earth with a large asteroid occurred. This event so radically altered earth's environment that the second largest mass extinction of flora and fauna, in earth's history, occurred. Luckily for mammals the dinosaurs were casualties of this great event. The end result of this was the lifting of a great predation pressure off the mammals and the concomitant opening up of seemingly endless niches in which to inhabit and multiply.

So why weren't mammalian species exterminated by this catastrophe? The answer is that some were but some weren't. The advantage mammals had over dinosaurs seemed to be two-fold. First, while mammals were *homeothermic* (able to regulate their body temperature) dinosaurs were thought to be *poikilothermic* (or incapable of internally regulating their body temperature). Much like lake turtles that find a flat rock in the sun in the morning to increase their body temperature, dinosaurs had to acquire similar strategies when cooler. When the "nuclear winter" generated by the asteroid impact persisted and persisted dinosaurs died out but, though affected, many species of mammals were successful at

1

surviving through the environmental disruption. Second, during the previous 150 million years some mammals had acquired a very beneficial *endocrine* (hormone) system. *Prolactin* and *DHEA* (*dehydroepiandrosterone*) were the two hormones thought to have aided most if not all mammalian species through the cold times. Both of these hormones had the ability to kick up the thermostat in mammalian bodies that would yield a significant advantage in combating cold weather.

All mammals continued to evolve through the process of natural selection during the next 60 million years or so to the point that the early hominids, *Australopithecus afarensis* and *africanus* appeared (in Africa). *Australopithecus afarensis* is thought to be most important as an ancestor of ours. *Homo habilis*, our more direct ancestor, appeared about 1.8 million years ago. It is thought that the more evolved species of *Australopithecus, A. robustus, A. boisei*, and/or *A. crassidens* may have coexisted with *Homo habilis*…*afarensis* and *africanus* having become extinct. Soon after, *robustus, crassidens* and *boisei* disappeared from the geologic record. *Homo habilis* is most significant in our evolution due to the fact that they appeared to be the first hominids to accumulate and transport stone artifacts. They also used animal skeletal parts for tools, thereby, it is thought, becoming the first true tool users and makers. *Homo habilis's* descendant, *Homo erectus*, is thought to be our (*Homo sapiens*) direct descendant and along with being the first to master fire, was the first to migrate into Eurasia.

Homo sapiens, having evolved from hunter-gatherer nomadic hominids, were, of course, also hunter-gatherers. Their average day revolved around acquiring, through hunting animals and gathering seeds, fruit, roots and leaves, enough food to subsist. Not to flourish but to *subsist*.

Occasionally, when things went well, they accumulated enough food that they could relax a bit but generally the selection pressure was that if they didn't totally focus, throughout their group, on acquiring food, they would starve. This selection pressure (they were selected against by dying if they didn't constantly forage) had been with all Hominids as well as all animals since the beginning of life. Eat or die. Critical and central to this theme is reproduction so really the instinctual rule is eat enough to subsist to the point you pass on progeny (reproduce) or you become extinct. This theme is the common fiber for all species on the earth at present. It was the central theme in all species that are no longer with us. Of all the species of plant and animal that has ever existed, less than 5% have representatives with us today. It is easy to get wrapped up in our own species but there is more to our world than just us.

Sugar

Now, we as humans, are not so far removed from being hunter-gatherers, say 11,000 years or so (based on the generally accepted fact that our earliest agrarian or farm-based, non-hunter-gatherer societies, started about 10,500 years ago in southwest Asia)…not to mention the 200 to 300 MILLION years of compulsory mammalian evolution through natural selective mechanisms before this, and all of a sudden we're "civilized". It's nice to think so. But let's take a seemingly innocuous topic like sugar consumption and see why it tastes so good and affects us like it does…lets see how civilized we are.

Why does sugar taste good? Ultrastructurally (looking at very tiny things with very powerful microscopes) it can be observed that there are sweet receptors located at the tip of the tongue (the tongue also has salt, sour, and bitter receptors) and to a lesser degree on the palate. These are basically "lock-key" triggers with a neural hookup directly to areas in the brain that evolved to appreciate sweet. When substances placed in the mouth break down to molecular entities they assume a three dimensional shape much like *Legos* or *Tonka* toys only very, very small and numerous. Sugars have a characteristic shape that fit in sweet receptors like a key in a lock, thereby activating the neural pathway to the brain and if there is enough perceived for enough time…viola!…you feel happy, excited, pacified or whatever. And as everyone knows it doesn't take much to activate the pathway and the response is usually the same…a pleasurable experience and a quest for more sugar.

The term "sugar" is a generic expression for a class of carbohydrates, usually crystalline, sweet, and water soluble. The sugar that everyone knows as "table sugar" or "cane sugar" is *sucrose*, a *disaccharide*. *Saccharum* is latin for sweet, appropriately enough, and so the term disaccharide means, of course, two sugars. Sucrose is made up of the two sugars, *glucose* and *fructose*. Glucose is a 6-carbon *aldose* (a chemical family made up of eight different sugars) and fructose is a 6-carbon *ketose* (a chemical family made up of four sugars). The way the two basic sugars are joined is the key in sucrose's value to consuming organisms. This basic bond is easily *hydrolyzed* or broken down, with energy derived from this. Then, the separate molecules can be further metabolized with their usual energy benefit.

3

This sugar (sucrose) is extremely abundant and hence extremely important in the plant world.

A 3-carbon aldose you have experience with is *glyceraldehyde* which is the main component of antifreeze. An important 5-carbon aldose is ribose which is a component of RNA and DNA (*ribo*nucleic acid and deoxy*ribo*nucleic acid), the "stuff of life". Lactose, *milk sugar*, is also a disaccharide, being comprised of two aldoses, one being glucose and the other being *galactose*. It is not found anywhere else in nature except in milk. *Maltose* is a disaccharide made up of two glucose molecules. Starch is all just glucose in 3-D linkage as in potatoes. Cellulose, the structural sugar of plants, is made up of glucose units (as parts of a repeating disaccharide called *cellobiose*) also but the individual sugar entities are linked in such a way as to make them difficult to break apart...hence their difficulty in digestion.

The problem with table sugar is that it just plain isn't good for you in anything beyond occasional exposure. We evolved with intense pleasurable responses to sugar simply because it often times meant the difference between surviving and not surviving. Our brain works on only one fuel and that is glucose...a component of sugar.

Let's look at the availability of sugar 10,000 years ago and before...it was very scarce in high concentrations. Aside from seasonal findings of ripe fruit and the occasional close encounter with honeybees and their honey, sugar was only around in very small, nearly undetectable amounts. This is why the taste is pleasurable...to make sure if we came upon some, say in ripening fruit, we would load up on it because it was important for survival. So important for survival, that our attraction to it was conserved throughout the entire population of humans that cover the planet. The other taste modalities of bitter, sour, and salt vary greatly in their importance throughout the cultures of the world but sweetness is pretty much the same in importance and intensity everywhere.

We take for granted sugar's existence today but even sugar beets weren't around in anywhere close to the form they're in today. It wasn't until the 18th century that, through an accelerated selection process, beets, with a higher sugar content were developed and used for sugar production. Sugarcane was only available once Europeans started their exploration and conquest of the seas, circa 1500 ad. Then only the very wealthy could afford this new source. Cane was later introduced to the Caribbean making sugar less expensive in the latter part of the 19th century.

It's very interesting that early Americans on the frontier had only maple syrup as a naturally occurring concentrated source of sugar. Honey bees, not naturally occurring in the Americas, were brought over around 1630 by colonists. Apples were a constant source of sugar and have come to represent America, but apples did not naturally occur in the New World. They actually first evolved in Kazakstan, north of Afghanistan, way back when, and branched out to Europeans and Asians from there. The Crab Apple naturally grew in America but it is a far cry from a regular sweet, ripe apple.

Sweetness has transcended its mere gustatory experience and throughout recorded history has come to symbolize the best or most coveted of things. Jonathan Swift wrote of "sweetness and light" indicating the highest ideal. Shakespeare described spring as "sweet o' the year". Michael Pollan, in his national bestseller *The Botany of Desire*, denotes sweetness as "a reality commensurate with human desire" standing for fulfillment.

So this is where we're at now. The number one crop in the entire world based on annual tonnage harvested, nearly surpassing number two (rice) and number three (wheat) *combined*, is...you guessed it—SUGARCANE! Why is this happening? As you can see the sugar industry is merely taking advantage of a deep-rooted evolutionary adaptation. One for survival that suited us well through 200 million years of natural selection perfecting it, only to not need it any longer because we are no longer hunter-gatherers trying to subsist. Rather we are members of an industrialized society that has enough food to feed everyone in excess!

The effect of this overindulgence is, if you ask the executives of the sugar producing and refining companies, wonderful, from an economic standpoint but negligible from a health-effects perspective. If you ask health care professionals and scientists not on the bankroll of the sugar companies the answer is "potentially devastating!!" Wow!...why the chasm between the two? Kind of reminds one of the Tobacco companies back in the sixties, doesn't it. It's nearly identical and just as hazardous in a benign sort of way.

This is what sugar means to us. When you have a birthday you get a cake loaded with sugar...on special occasions like Christmas and Thanksgiving you have dishes and desserts loaded with sugar. Think about how deeply rooted sugar is in the special moments in your life and it's not so hard to see why it's in most "comfort" foods. It's in all the specialty coffees, it's in bread, heck, it's in just about everything processed. Psychologically we need it because we are so used to it. Throw in Wall Street advertising to make you think that you're not thinking right if you don't drink or eat it and...and all because of some mutation long ago that conferred extra survivability to those perceiving sweet the way we do.

The effect of this misuse is that we have more obese people, many, many, more Mature-onset Diabetics (NIDDM), an incredibly high need for dental care, chromium depleted over users, immune-suppressed over users, thiamine deficient over users and the list goes on and on.

Recently researchers on the aging process studying different ways to prolong life stumbled upon an incredible finding. Caloric intake consistent with *subsistence only*, led to a 40% extension in lifespan of lab animals!! It sounds like if you conduct your eating practices more in line with our evolutionary ancestors that you may be more healthy and long-lived. Makes sense doesn't it. When's the last time you saw an obese 80 year old? There aren't many. Look at people that lived a long time in history. Marie Aouret (Voltaire) was a subsistence eater and lived to be 86 years old in the 1700's! More incredibly Sidhartha Gautama (the Buddha) was another subsistence "exister" who lived to about the same age only this was way back circa 500 BC! Two people certainly doesn't prove anything but its all kind of making sense.

The basic axiom to this that will be revisited with other topics is that evolution through natural selection has fitted us with survival mechanisms that have become outdated due to our overabundance of time and resources. Because we live in a capitalist society we are being preyed upon by people who care only about making money who are willing and able to exploit this obsolete urge.

Fats

Fatty Acids have played and continue to play an extremely important role as an energy rich fuel in plants and higher animals. The main reason for this is two-fold. First, fatty acids are high in energy, yielding 9 kcal/gram compared to carbohydrates (4 kcal/g, protein 4 kcal/g, alcohols 7 kcal/g). Second, fat can be stored in cells in a super-efficient nearly anhydrous (not coupled with water) form, as fat droplets, making fats the quintessential weapon at combating food shortages. Oddly enough, though, free fatty acids are somewhat toxic, which means they must be tightly bound to protein when transported in the blood.

When used as fuel a common fatty acid like say *palmitic acid* (a 16 carbon saturated fatty acid common in animals) undergoes a complex activation cycle after which it is transported across mitochondrial (each cell's powerhouse) membranes into the mitochondria's main compartment. Here it is degraded, 2 carbon sections at a time, via the "fatty acid oxidation spiral" yielding lots of acetyl-CoA and *lots of hydrogen atoms*. The 2 carbon acetyl-CoA sections are then used to feed the *tricarboxylic acid cycle* (energy yield) and the resulting hydrogen atoms enter the respiratory chain as electrons, to pass to molecular oxygen via the *cytochrome system*, allowing *oxidative phosphorylation* of ADP to ATP to occur yielding transportable energy (big energy yield). Lots of somewhat complicated stuff but the important part of all this is the energy yield, especially from the hydrogen atoms. This is why saturated fats are not as good for you as unsaturated fats or polyunsaturated fats…because what they're saturated with is *hydrogen* which is converted immediately to energy and stored in the form of ATP (adenosine triphosphate, the cell's energy currency). Animal fats tend generally to be saturated with hydrogen and vegetable fats tend to be unsaturated which means not as many hydrogens which means less energy.

So…if you were a hunter-gatherer on a subsistence diet living day to day and you came upon animal fat in conjunction with protein you would be selected for to load up on fat due to its high energy content. Due to this evolutionary predisposition toward fat we have acquired a liking for the taste and feel of it much like we acquired a taste for sugar. Unfortunately, these days with activity levels down and calorie levels up you would do better to really, really moderate your intake of animal fats or be willing to pay the price. This is why dehydrogenated fats, unsaturated fats, and vegetable oils are recommended by dieticians and physicians over their saturated cousins. It's not that they're good for you...it's just that they're not as bad. Of course, this is also why fats in general and saturated fats in particular are appealing to us. Look at some of the food loaded with fat: chocolate, gravies, bacon and sausage, sauces, food spreads, creams and other dairy products and the list goes on and on.

The way the marketplace is with advertising in regards to fat is interesting, though. Several decades ago the saturated versus unsaturated gambit was used by companies to make sure you continue to buy fats but hopefully more of the unsaturated ones. Still, fats pervade our kitchens and restaurants to such a degree that you really have to be vigilant to get just the amount you require (which ain't much). So even though fats make foods more pleasing and make them slide down the old chute easier...you still must really moderate.

Smut

Talk about exploitation of an obsolete urge, the purveyors of smut (including Playboy and all related magazines, internet sites dealing in porn, prostitution and most bride import businesses) take the cake. Of course the urge to procreate is not really obsolete since we still have to pass on progeny, but the intensity of the urge could probably be turned down a bit.

Hundreds of thousands of years ago with subsistence living being all the rage (like there was a choice) it was absolutely necessary for humans to place paramount importance on reproduction. If you go further back to our ancestors the need was just as great. Our species, as with any species, had to be sure that enough numbers were carried forward.

The problem with "long ago" is that there were no hospitals, no doctors, (no lawyers, which was good) lots of predators and though not nearly as many as today, some diseases. Consistent food sources were iffy so humans had to keep doing that hunter-gatherer thing. Let's face it, the whole picture was not conducive to safely and consistently giving birth and rearing younger children. The rates of infant mortality must have been very high. So obviously couples couldn't plan to have just one baby if they wanted a family. Go back far enough and there was no planning involved, the intense urge based on self gratification through a high amount of excitement and pleasure had to be enough (and obviously it was). Our ancestors had to keep turning babies out and hope that half or more made it through to the reproductive years. So even though the men were dead tired from all the hunting, gathering etc that they did on a daily basis, on their subsistence caloric level, they still had to have enough of an urge to copulate whenever and wherever possible. The women, after all they did on a daily basis, still had to have an urge to engage in sexual activity also, though because their investment in the pregnancy was 100%, unlike males, they were selected for being not so indiscriminate and thereby not so seemingly mindless about it.

Fast forward to today's society and since, in evolutionary terms, we again are not so far removed from the eons of natural selection that got us here, its not too difficult to see how easy it is to separate men from their money when it comes to anything involving the opposite sex. Make it available in any way and men will come running. Do the opposite and make male smut available to women and the response is not nearly as intense…and why would it be, based on our evolutionary history.

So the axiom to be revisited here is: evolution through natural selection has fitted us with a survival mechanism that has become outdated through time as it regards the need to over procreate. Because we live in a capitalist society men are being preyed upon (women to a much lesser extent) by people who care only about making money.

What we do about this "dependence" on it is a difficult question. It has become nearly a cultural obsession. The most important thought in regards to this, I suppose, is *moderation*. It's OK to enjoy a good amount but don't get carried away. And certainly don't engage in high risk sexual activity that invites disease or worse.

Greed

It is thought that in the animal kingdom there are three basic variants of individual-individual interaction. They are the *reciprocal* approach, the *altruistic* approach and the *selfish* approach.

Altruism (expending energy for nothing in return) is genetically a dead end strategy except in cases of *kinship*. Obviously if you were an organism running around devoting all or even a modest portion of your time and energy to enhance another *unrelated* organism's genetic fitness to the detriment of your own, well its not too hard to see how your genes may not persist too long in the gene pool. So its one thing to be a philanthropist and donate sums of money you can easily do without for the purpose of feeling good about yourself and acquiring prestige but quite another to be living day to day, barely making it and dedicating time or resources to another for no benefit whatsoever. You might make it short term if you're lucky, but in the long term there's no hope. Altruistic organisms can't be found because it's a flawed evolutionary strategy.

The reciprocal approach on the other hand, tends to be evenly distributed throughout the human population and an amazing array of other species. In other species, where there are no higher thought processes, they refer to it as *symbiosis*. Symbiosis is further subdivided into *commensalism*, *mutualism*, and *parasitism*.

There are incredibly interesting examples of symbiosis. The *Fierasfer* is a small commensalist fish that lives in the rectum (yikes) of the sea cucumber. This little animal periodically emerges to feed outside the cucumber but returns to safety by poking the rectal opening with its snout and then quickly turning around so it will be drawn tail first back inside the cucumber's rectal chamber.

Another example is that of the tiny crab that lives in the mantle cavity of oysters. The crab enters the cavity as a larva and eventually grows too big to escape between the two valves of the oyster's shell. It steals food in small quantities from the oyster and is protected but otherwise does no harm.

There are many other examples. Symbiotic mutualists are numerous...birds and bees, legumes and nitrogen-fixing bacteria, and termites or cows and the cellulose-digesting microorganisms in their digestive tract to name but a few.

Then there are the parasites. Some examples briefly are mosquitoes, lice, tapeworms, protozoa, viruses and bacteria.

Reciprocal behavior in humans is very common and well documented with even some sayings having evolved along with the strategy…"you wash my back I'll wash yours, tit for tat, eye for an eye"…give it a moment and I'm sure you can come up with many examples of sayings or occurrences where something was done or given because of a previous gift or act.

This brings us to the selfish approach. Avarice, selfishness and greed…why do they persist in the human population and what are their implications? In bee colonies, termite colonies, and ant colonies you will not find any of this. This is because the general rule of greed and selfishness is that the organism exercising the selfish act is decreasing the genetic fitness of an unrelated organism while enhancing his own fitness. In bee, termite, and ant colonies all members are closely related. In human populations greed is endemic (even among related individuals). Competition and a "no holds barred" approach in business is mainstream. It's competition at its keenest. Even in day to day interaction selfishness is common. When's the last time you sat waiting for a parking place with your blinker on only to have it taken by someone who most probably knew you were waiting but didn't care. It's so common that we're used to it. We may not like it but we have seen it enough that we're used to it. It's especially common when dealing with people while driving because of the anonymity afforded by the feeling of isolation when inside a vehicle. People are much less likely to be selfish when the other person is known.

It's easy to see why or how selfishness persisted in human populations during the hunter-gatherer times. Competition between rival groups was so keen just to survive that anything went. Selfishness inside a hunter-gatherer troupe was surely much less common.

Today the 1.8 million years of natural selection that encompasses recent hominid evolution has yielded many different strategies of getting what one wants. In order to survive to old age and not have the strategy you adopt affect you deleteriously, one must pick when to be selfish so as not to offend the larger or more offensive minded person or group. Even in situations where one might think one has a "pigeon" to take advantage of, one never knows because of the unknown crazy "overreactor" that may reach for a gun at the slightest provocation. So my advice to you is to live within the confines of one of the oldest laws on or off the book which is "Do unto others as you would have them do unto you".

Exercise and Activity

When studying any mammalian body *in vivo*, from the tiniest mole to the largest whale, it is very evident that activity level is directly associated with the organism's health. Obviously there should be latitude when interpreting activity levels since animals must sleep and some must even hibernate but, that aside, when others of your species are active then you are active (or you're too young to care for yourself or too old to care for yourself or you're injured). Humans and the animals they've domesticated and spoiled are the only mammals that actually, en masse, evade activity. This unfortunately is much to their detriment.

It has, in the last century, become vogue, to be lazy. Starting with our Industrial Age of the late 19th century and the concomitant wealth it bestowed on certain individuals, the media (it's always the media's fault) has covered the "lifestyles of the rich and famous". People read about these individuals and later (by way of radio and television) as the media became more sophisticated, actually were able to see for themselves, that groups who were successful had, as their reward, the luxury of doing nothing except lounging around and the like. Average people tend to idealize the lives of people that are rich and famous. The goal of too many people in their lives is to be progressively more sedentary, as a way of making themselves feel better, by being like the idealized persons they've perceived. It's no wonder that, in conjunction with the average person's estrangement from a healthier subsistence diet, this approach to activity is leading to deadly results. Mix low activity levels with over consumption of sugars and fats and it's very easy to see why the incidence of NIDDM (mature onset diabetes), colon cancer with its predisposing conditions diverticulitis, diverticulosis, and chronic IBS (irritable bowel syndrome), DVT (deep vein thrombosis) as well as the myriad of coronary artery diseases and other circulatory problems, not to mention the endless list of cancers, pervade our population.

It truly is a tribute to modern medicine that we can actually have an average life expectancy that is steadily on the rise in spite of the crazy ways we care for ourselves. Unfortunately the flip side of that is that people tend to become over reliant on artificial pharmaceutical means at recovering from maladies or just not feeling like they want to. In watching an Ellen Degeneres standup comedy bit the other day I was struck by the poignancy of a segment of her routine. In it she was mocking the media especially commercials that were attempting to prey on human insecurities as they relate to the psychological health of the viewer. She says in the third person, as if spoken by an overly disingenuous 'salesperson', "are you anxious, are you depressed?" and immediately after she says in her own declaratory delivery "yes...I'm human". How true this is. We all, as humans, have times when we feel down or are anxious. The funny thing about it is that, aside from a small percentage of frank psychological disorders, were making our own bed by an incorrect approach to what we put in our bodies and mishandling the energy reserves we naturally have at our disposal. Does anyone think human hunter-gatherers had no depression or anxiety? Of course they did but they were too busy trying to survive to let it consume them as so many do today. There's an old saying that states that the best treatment for the blues is industry. Getting to work on something that, when you're finished, you can stand back and feel good about what you've done. The roots of this are obviously in our past when we didn't have so much time to sit and contemplate how hard life is on us.

Sleep

Like so many biological aspects of members of the industrialized world, sleep is more often than not looked at as something to work around and begrudgingly do as opposed to being something of critical importance that needs to be a priority. It's as if given all these instinctual cues the people who can ignore the most and still "maintain" are the ones that are revered. Think back to younger days if you're older or just life if you're 15 years to about 28. People brag up their 'ability' to go without sleep and 'party hardy' and are actually respected for their deviation from what is required. Is this a surprise? Hardly…just look at the media and how it plays up deviation. The further events deviate from the norm the more likely it is they will make the news.

We live in a "Guiness Book of Records society". Nobody wants to hear the average or mundane…they want Evil Knievel jumping the Grand Canyon and breaking all the bones in his body. They want "how many goldfish the guy swallowed". They certainly don't want to know how John Doe did thing "average" on day "usual". It's the fringe of the Bell Shaped Curve, baby! The Gaussian distribution statistically assigns where a particular trait or behavior or occurrence is in relation to the norm. People, events, traits or occurrences further from the norm are accompanied by fewer others until you get to the fringe where there are few because the trait or behavior or whatever is so abnormal or difficult or deleterious. For a simple example, most people "need" 7.5 hours of sleep. Eighty percent of the population falls in this area of the bell. Of the twenty percent remaining, ten percent need more and ten percent need less sleep. Of this ten percent, there are the two percent either way that are way out from the norm, requiring 4 hours or conversely needing 12 hours. Further from the norm than this and either the brain doesn't function properly from lack of rejuvenation time or you have serious Thyroid or Pineal gland problems.

How much sleep as an individual do you need? It's not too difficult to find out. You just have to keep a journal and be able to do some basic math. It's probably about 8 hours like most people, but find out. Now for the stuff about sleep that you may not know.

Prior to attending a lecture on sleep last year which presented findings from the Harvard University Sleep Symposium I was bothered by the fact that I often woke up three to four times during my eight hours of sleep at night. Rarely, if ever, did I sleep all the way through the night without stirring. After waking in the middle of the night, it wasn't unusual, if I had to work the next day, to feel the "pressure" to need to get back to sleep so I would be 100%. Unfortunately, this often worked in reverse which kept me from falling back asleep in a timely fashion. My sleep partner, my lovely wife Joan, it seems has a nocturnal tendency called "restless leg syndrome". She in turn, insists that I have "cricket foot" or a tendency to rub my feet together while asleep. This causes her to wake up occasionally also. The odd thing is that when she goes to her girlfriend's for a sleepover every 6 months or so, I find myself waking even though she's not there to wake me. As it turns out, we as humans are programmed to be alert, even at night, though only periodically. We don't just drift off to sleep, bask in la-la land for x amount of hours and when the brain buzzer goes off, wake up. You knew it would be more complex than that, didn't you?

What happens is that needs are prioritized…body versus brain. Since we are products of evolution through natural selection, and brains, especially the cerebrum and neo-cortex, are relatively recent developments (relative to our muscular or not so muscular bodies), the priority of rejuvenation through sleep is yielded to the body. Yes, the body is rested, reworked and enzymatically and chemically reoutfitted first, prior to the brain. The brain later is given several opportunities to chemically and electrically reset and if the proper amount of rest is obtained the individual is ready to go, both mentally and physically, after say 6–8 hours. Here's basically how it works.

Amazingly everything revolves around a 90 minute timer. This is apparently a highly conserved strategy since all *Homo sapiens* that are not beleagured by brain pathology or influenced by drugs are governed by this cycle. This means that it developed early in hominid evolution and was important enough to be conserved or retained to be widespread in the human population of the earth!

What happens when we fall asleep is that we slide down into our deepest sleep to allow our physical being to recharge. We stay in this state up to 70 minutes or so and, though nearly unarouseable, if aroused we can be most unpleasant and unpredictable. During this time we dream our wildest and strangest dreams yet almost never remember them. We slowly ascend out of that stage to a step closer to wakefulness yet still we're out of it, pausing briefly heading closer and closer to REM (rapid eye movement) sleep through still another stage and finally we are there. In REM sleep we are as close to being awake as possible without actually being so and are easily aroused to wakefulness. If you have to void your bladder or hear a noise or feel movement it's now you will wake. It's also in this all important REM stage that your brain undergoes its biochemical and electrochemical revitalization. This is also the time in which you dream most life-like or real dreams.

Dreams occurring during REM sleep are almost invariably the dreams that are remembered, if the individual remembers any.

The first REM stage doesn't last but a few minutes and if not aroused to wakefulness the brain descends back down into the deepest stage once again. That first cycle, having lasted about 90 minutes, replenished the body mostly and just yielded a little brain rejuvenation. In the second 90 minute cycle the pattern is repeated with the difference being that the deepest sleep, when the body is recharged more, is much less long, and the REM stage is appreciably longer. As subsequent cycles occur, REM sleep dominates, with but a brief stay in the deepest sleep stage in each cycle. Finally after 7–9 hours on average, you arise, hopefully refreshed and ready to go (if you wake out of REM sleep you will be more refreshed and ready to go than if you woke out of a deeper stage).

The 90 minute intervals are amazing, though. I find myself going to bed at 10PM generally, and I will wake up for some reason at 1:12AM or very close to that every night. Pay attention to when you wake up next time. Allow for the time it takes you to fall asleep and note that it will most likely be a multiple of 90 minutes.

As humans age into our 7th and 8th decades of life this pattern is generally maintained with the exception that the deepest sleep is not experienced very often any more. More often than not the brain only slips to the stage just before the deepest sleep but not quite into the stage where the body revitalizes itself best. Also, the periods or intervals, at each stage, are inconsistent. The trend is that sleep doesn't get better with old age, unfortunately (although studies show that regular exercise in fit older people allow more deep sleep episodes). REM sleep occurs but only briefly usually and the 90 minute intervals, again, are not so consistent.

Medication or recreational drugs and alcohol have a negative affect on the normal rhythm of sleep causing a disturbance on the depth or length of various stages depending on the dosage and the mix.

So what does this mean? Respect your bodies' need for sleep and find out what it takes to improve in this aspect of your life…you won't be sorry.

Circadian Rhythm

What's really interesting and of great importance in aligning one with one's evolutionary self (in order to have both your body and mind "purr like a kitten", so to speak) is your body's circadian rhythm. *Circa* is latin for "about" and *dies* means "day". So circadian rhythm is the rhythm of our bodies as they relate to the 24 hour period we call a day (1 of our earth's rotations as it travels around the sun).

Both the plant and animal kingdoms are overfilled with examples of the pervasiveness of the rhythm and how critical it is to the species involved. For example, there are many flowers that open and close at a particular time of the day to ensure pollination by the appropriate insect. Leaves of many plants show regular movement in a cycle of approximately 24 hours. Many species of crabs turn darker in the morning and lighter later in the day to help camouflage them. Bees are especially amazing due to the fact that they can do a wagging dance in the darkness of their hive that tells other hive members the exact location of a food source. As the day progresses they can change their dance to compensate for the movement of the sun. This 15 degree shift per hour compensation is rooted in the bee's internal clock. "Crowding in time" is a predator evasion strategy that is used by many, many species. Pregnant wildebeests living in large herds, for example, give birth all at once to create a glut of food for predators so that most can survive the first critical week of life. Crowding in time is also manifested in *en masse* exits of cave crickets, cave bats, oilbirds, swallows, and other animals that take communal refuge in shelters. By creating confusion by their large numbers leaving a cave all at once they are selected for to have this behavior persist. All these are examples of the importance of circadian rhythms on animal populations.

These rhythms are just as important to humans. Incredibly we are very much creatures of habit and our approach to day to day life should keep this in mind. *Uncoordinated phase shifts of the various biological clocks in the same individual may well lead to serious physiological disturbances and perhaps to disease.* That sounded important so what does it really mean? Amazingly, different organ systems within the same individual can have different clocks. They still revolve around the 24 hour main-clock of the individual but since *humoral* (like hormones floating around in the liquid of the body) control of organ systems is so important, a regular necessary release of a hormone at a certain time of day or a certain day of the month can be disrupted by throwing off you clock by "jet-lag" or by staying up much too late, getting up too early, or worse yet by a combination of the two. Add to this a disruption in nutrient intake and a little stress and you can really mess things up temporarily. This is why you may notice or can remember a time when you went out of your routine only to find yourself coming down with a cold or the flu as a result. This is why consistency in one's routine, especially in one's sleep pattern is so important.

Now if you're sitting there saying that, years ago, as a youth you lived a disrupted lifestyle constantly or something of that nature, it's probably true. The human body is a remarkable entity and can take a fair amount of disruption. Look at what humans have survived in the past…the Jews and others in concentration camps, soldiers on the Bataan death march, and Civil War soldiers at Andersonville to name just a few examples. Survivors of these events often times went on to live out normal lives after their recovery.

Like I said, the body is impressive in it's ability to adapt. But the point is would you rather swim up river or with the current? Would you rather tack into the wind or sail with it? When you want to swim as fast as possible in a river you pick the right direction. When you want to do anything in order to maximize performance you try to get as close to optimal conditions as possible. This is just common sense. So why not align your body with optimal conditions for best performance. Find out what the optimal operating rhythm was historically as we evolved and adhere to it. This is not asking a lot. It means respecting your sleep period by not deviating more than two hours from it unless you absolutely have to. Eat regular meals at regular intervals. Get into routines of diet and exercise that are consistent. Then stick to them. Things like these are much more important as we age but even if you're younger and are an athlete or in school with a heavy load, to optimize your efforts you should do this too.

Athletics and Learning

I have been involved in athletics and academics all my life. I attended high school and graduated. I attended a Junior College and graduated. I attended a four year school and graduated. I attended Grad school and graduated. I helped to raise five children and observed and directed them through their educational process (as much as I was allowed to). I played organized baseball, basketball, and football. I swam for my high school and swam (intramurals) in college. I was a *karateka* at my dojo under a *sensei* for several years and represented my dojo in karate tournaments. I taught karate at my college. I've competed for my high school in golf and continued to golf competitively after high school. I've fenced, played foosball tournaments, played soccer, competed at trap shooting, played racquetball tournaments for years around my state, bowled, water skied and snow skied, hunted and fished. I've play badminton, tennis and squash. I've climbed a mountain, gone camping and hiking and shot bow and arrow and hunted with it.

I really have been involved in athletic and academic competition all my life. I enjoy the quest for excellence. I know what many of you are thinking and I agree…often times its not really how you do but merely participating is what's important. I have absolutely no problem with this. It is always better to be involved and active rather than to be inactive sitting on the sidelines with television watching as the center of one's world.

What I see all too often, though, is participants getting upset at their performance in a particular activity, yet they are unwilling to put in the extra time practicing to upgrade their performance level. To improve at anything takes practice. To get exceedingly good at anything takes dedication and consistent effort. The one advantage people who excel use, that the average person doesn't seem to understand (many have been seen to understand this all important rule especially at higher levels of competition), is the approach of consistency. It doesn't matter if we're talking about improving at something physical or something mental. The way your body or mind improves or understands better is to see or do things regularly. What it doesn't like is inconsistency.

Unfortunately most athletes and students don't do anything consistently. What they do is procrastinate. It's almost the rule rather than the exception…procrastinate until the night before the big test and then cram. Or worse yet, come into the important athletic season knowing you should've done more to prepare and get injured due to this lack of preparation. What your body wants or what your mind wants is to consistently see what you're going to be doing and then it will allow your best effort.

The backbone to this approach is *enzyme induction.* A simple example of this is alcohol (ethanol) tolerance. I will assume that everyone has experience with alcohol intoxication (for those of you who don't have any experience here, good job, keep it up, and follow along). When you were younger and you took your first drinks, you felt the effects of alcohol quickly. As you got older and drank more and more consistently, more alcohol was required to achieve the same affect. Some people who imbibe heavily over years and years can consume large amounts of alcohol at one sitting, an amount that would drop a non-drinker cold, yet they are only moderately impaired. The reason for this phenomenon is enzyme induction.

Alcohol is perceived by the body as being a poison and so when it is absorbed into the blood, the liver (the body's detoxification organ) is stimulated to produce *alcohol reductases* and *dehydrogenases.* These are enzymes that cause the break-down of the alcohol molecule. The more consistently you drink, the larger the area in the liver is recruited to make the enzymes, the more efficient you become at breaking down alcohol and the more you require to achieve the same level of inebriation (interestingly a side affect of chronic drinking is the enlargement of the liver and also the spleen, a secondary detoxifier, the liver also is a storage site for fat and sugar, and since alcohol is rich in chemical energy, this storage contributes to its enlargement).

Alcohol detoxification is a great example of enzyme induction yet not really unique. Weight-lifters experience this phenomenon when training in a very measurable way…the weight they progressively are able to lift (and the size of their muscles). Swimmers measure their induction of a myriad of enzymes through faster times in their events. It just goes on and on.

Learning and brain power operate the same way, basically. If you ever want to learn something well, expose yourself to it daily, for a half hour, over a three week period and you will not forget it. Cramming 10 hours in a row studying the same material may yield a satisfactory test score but the information will perish more quickly from your memory. So as a student, be efficient, be consistent, and you will learn much more effectively.

It's not too hard to understand. We evolved through natural selection to respond to cues consistently in our environment. We are creatures of habit because our internal organs are most efficient when our clocks are not disrupted. When forced to respond to unusual stimuli, we as a species offered a broad range of reactions. Many of these reactions were disruptive to us as individuals causing morbidity and reproductive effects. Through sheer numbers a certain reaction to any new stimuli became standardized through genetic promulgation and homeostasis ensued.

CHNOPS and the Cycles of Life

Ask the average person what *chnops* is and phonetically they interpret it as a european liquor. Ask a scientist of life processes and one finds a completely different reply/meaning. By taking the first letter of the six elements most critical to life, this word is formed. **C**arbon, **H**ydrogen, **N**itrogen, **O**xygen, **P**hosphorous and **S**ulfur yield **CHNOPS**. The importance of these elements cannot be overstated. Without any one of these elements, living things cease to exist. There are 92 naturally occurring elements on our planet and although the six elements mentioned above carry special significance it shouldn't be misconstrued that these are all that are necessary. Other elements of great importance are calcium, sodium, potassium, magnesium and iron. Also bearing consideration of importance, though limited, are the elements boron, fluorine, chlorine, manganese, cobalt, copper, zinc, selenium, molybdenum and iodine. The aforementioned 6 elements, though, are the subjects of my discussion. *Nature's Building Blocks*, by John Emsley, was an inspiration to this section and was relied on heavily for various parts.

Sulfur

It's no accident that the material of life, the stuff of reproduction and inheritance, the biochemical deoxyribonucleic acid (DNA) is comprised of five of these six most important elements. The only one of these six not involved, sulfur, is so important as a component of an essential amino acid (methionine) specifically, and in a life giving cycle generally, that it makes sense to defer it special consideration.

Sulfur is the 16th element of the periodic table and occurs naturally in 4 different forms, called isotopes. None are radioactive and the most common is sulfur-32, comprising 95% of all the earth's sulfur (sulfur-33, sulfur-34, and sulfur-36 are the others). The "32" denotes the weight or "atomic mass" (designated in *Daltons* or *atomic mass units* which is one-twelfth of the mass of a carbon atom) and indicates there are 16 protons and 16 neutrons in the nucleus of each sulfur atom. Sulfur-33 has 1 more neutron but the same number of protons. Sulfur also has 16 electrons buzzing around the nucleus (where the protons and neutrons are) and although these electrons can vary in distance from the center of each atom, the <u>number</u> of electrons, in an *uncharged* sulfur atom, will match the number of protons (which for sulfur will always be 16).

The distance from the nucleus electrons "reside" give rise to the reactivity of the atom, or "desire to interact", of especially the electrons furthest away from the nucleus. Sulfur, because of what chemists call "valences" and "orbitals", happens to have 6 electrons in its "outer shell", but would much rather have 8. This would satisfy what is called the "octet rule" of stability. Because of this, sulfur loves to interact with many different elements. It is in the same family as oxygen and selenium and therefore shares many things in common with them, all because of this "level of reactiveness", 6 out of 8 electrons in its outer shell.

The opinion has been repeatedly voiced by scientists that, much like an oxygen-based life form dominates here on earth, elsewhere in the universe there is almost assuredly a sulfur based life form dominating a planet. Earth does in fact have life forms that depend on sulfur for energy just as we depend on oxygen. One, called *Pyrobaculum islandicum*, lives now around hot springs and lived long ago during the Permian and Jurassic Periods and was responsible for depositing large amounts of uncombined sulfur which provided sulfur for a wide array of sulfur-loving bacteria.

Sulfur has been known to man for centuries. A name you may recognize it by is *brimstone*. It has been used in warfare as a component of "greekfire" (a flaming concoction hurled by the Greeks at their enemies), as a medicament for such varied ailments as indigestion and acne to parasitic infections and constipation, and as a fumigant and incense. Today it is in widespread use in medicines, foods, and industry.

Rotting eggs smell terrible because on the decomposition of sulfur-containing complexes. Onions and garlic have their characteristic odors because of di-2 propenyldisulfide and thiopropanal S-oxide, respectively. Lamb, passion fruit, roasted coffee and peanuts all derive their characteristic smells from the presence of sulfur. Vitamin B1 or thiamine, has sulfur in it as does the essential compound biotin. It is also in widespread use as a fertilizer in farming. Sulfur truly is essential to our lives, which is why understanding its natural cycle in our huge ecosystem called earth is so important.

There are three main compartments of sulfur in the environment:

1. on land: there are approximately 3 billion tons as sulfate (combined with oxygen usually in salt form) and about 600 million tons in living things

2. in the seas: roughly 1.3 *million billion* tons as dissolved sulfate and 24 million tons in living things

3. in the atmosphere: around 3.5 billion tons of gaseous sulfur compounds

Most organisms take in sulfur as sulfate (like your lawn does when you put ammonium sulfate on it). They then reduce it (add a proton) so it can be incorporated into the amino acids methionine and *cysteine* (another sulfur-containing amino acid though not as important as methionine) needed for protein synthesis. Marine algae (think of how big the oceans are and how much algae there must be) take in sulfate, make cysteine and methionine, but convert most of it to *dimethylsulfonium propionate*. This is what they use to maintain their osmotic balance with the salty sea water. When algae die or are eaten this compound is converted to *dimethyl sulfide*. If you frequent ocean beaches, this smell is very familiar to you...especially in the late fall. Dimethyl sulfide is a gas and it is in this manner that sulfur is released into the atmosphere (transferring as much as 50 tons to the air per year). In the atmosphere dimethyl sulfide is changed to *dimethyl sulfonic acid* by reacting with free radicals and then changes chemically to *sulfur dioxide* and eventually to *sulfuric acid*. It appears that this reservoir of sulfuric acid acts as our planets temperature regulator by promoting cloud formation and thus helping to cool the planet.

You've most likely heard of sulfuric acid and sulfur dioxide (both are released as byproducts and waste products of industry and are components of smog). What we are doing by adding to these sinks artificially is contributing to the planet being out of balance chemically. Acid rain was a catch phrase back in the eighties and pollution control became a priority of the EPA (environmental protection agency). They did some good but not enough. Acid rain continues to be a problem not only domestically but especially abroad in countries that are in a hurry to expand their country's industrial complex but are too cash strapped to concern themselves with the side effects of their endeavors.

The problem with acid rain is that it disrupts the sulfur cycle. Too much sulfur dioxide (a waste product of fossil fuel burning as well as the industrial complex) leads to too much sulfuric acid which can then fall to the ground as acid rain. Acid rain causes nutrients to wash away from plants and it changes the pH of the soil which allows other pollutants like aluminum to become free. Aluminum normally is not present in the soil in appreciable quantities but in polluted areas can displace biochemically important elements like magnesium and calcium. Too much sulfur dioxide also can choke off the ability of plants to take up carbon dioxide by blocking their stomata.

This is all very serious and with the cycles in nature involving the other components of CHNOPS you will see how we are really throwing our planet out of chemical balance.

Carbon

By discussing carbon second, behind sulfur, I in no way meant to relegate it to a less important role than it has on our planet. There is no element more important to life. Chemistry devoted an entire branch called *Organic chemistry* to it. It is a part of nearly everything we do, eat, and are. Even DNA, the "stuff of life", has its main component of carbon. The reason for its importance is simply that carbon is capable of forming strong single bonds, with itself, that are stable enough to resist chemical attack under usual conditions. This allows carbon to form rings and chains that are the structural basis for many compounds of living cells. As discussed in the sulfur section, the outside shell electrons determine the "need to react". Carbon has an atomic number of 6 (6 protons and six electrons) and has 4 of the 6 electrons in its outer shell. According to the "octet rule" it wants eight, so it is really ready to share its electrons with other elements and even with other carbon atoms. It is the master of the *covalent bond*.

Central to carbon chemistry and life is the process called *photosynthesis*. This is a metabolic process that is fundamental to nearly all living organisms, because of its occurrence in plants. Essentially, what occurs during photosynthesis is this: light absorbing pigments (*chlorophyll*) capture light energy from the sun and convert it to chemical energy, ultimately in the form of glucose (a 6 carbon molecule, mentioned previously in the Sugar section, that has oxygen and hydrogen in it too). *Carbon dioxide* from the air is consumed in the reaction as are the protons (hydrogen) of water, thereby releasing oxygen.

As you might guess it is a tad more complex than this with both a light phase and dark phase during which NADPH (nicotinamide adenine dinucleotide phosphate) is used to rob water of its protons (reduction) and later oxidized back to NADP thereby releasing energy that is used to make carbohydrate. Simultaneously ADP (adenosine diphosphate) also behaves similarly, gaining a phosphate (stored energy) and releasing it during which time an enzyme catalyzes the reaction forming carbon bonds for carbohydrate (sorry, I had to elucidate).

The historical work on photosynthesis by an English clergyman and chemist named Joseph Priestly in 1771 was brilliant as was the contributions of Dutch physician Jan Ingen-Housz. If you look into the works of these people you will be amazed at how little they knew back then and how much they accomplished with so little they had to work with. At any rate, solar energy captured by the process of photosynthesis is the source of well over 90 percent of all the energy used by man for heat, light, and power (since coal, petroleum, and natural gas, the fuels used by most machines, are all decomposition products and byproducts of biological material generated millions of years ago by photosynthetic organisms).

Many people believe that photosynthesis is only carried out by more advanced plants but I'm here to tell you that it ain't so. Actually more than half of all photosynthesis takes place in the oceans by small plants called phytoplankton. Even more reason to respect the environment and the carbon cycle.

The big daddy of all the cycles on earth is the Carbon cycle. This cycle rules the tempo of life on our planet while turning over 200 billion tons of carbon each year. There are 5 main compartments of carbon in our ecosphere:

1. Atmospheric carbon in the form of carbon dioxide, carbon monoxide, etc., comprises about 724 billion tons.

2. Living things on land form a reservoir for about 2000 billion tons of carbon.

3. There are 39,000 billion tons of carbon in the oceans mainly as dissolved carbonate.

4. 40 billion tons reside in living things in the seas.

5. In the earth's crust there are 100 million billion tons in the form of carbonate rock and 375,000 billion tons as reduced carbon (this includes coal, oil, and gas).

The absolute ruler of biochemical processes on earth which allows for nearly all other life and is central to the carbon cycle is the process previously discussed, called photosynthesis. This allows for the first step in the movement of carbon into plants which then provides food for animals. In this manner, carbon is passed up the food chain with each participating organism releasing some carbon dioxide until most carbon is back where it started. Quite an amount though is deposited on land and is eaten by microbes and other creatures in the soil. Similar events happen in the oceans with the difference that much of the carbon ends up on the ocean floor as carbonates where they can remain indefinitely.

Carbon, in the form of carbon dioxide, is critical to our ecosphere. Much like sulfur acts to a certain degree as our planetary temperature regulator by forming sulfuric acid in the atmosphere which causes cloud formation, carbon contributes in a big way also. Heat radiated from the earth is absorbed by carbon dioxide and radiated back to the earth's surface. A little is good but too much is not good and contributes to throwing our environment out of equilibrium (*Greenhouse effect*). By releasing all the carbon dioxide we do from burning fossil fuels and destroying so much of the earth's vegetation we're compromising our earth's ability to maintain this equilibrium. Remember, plants and chief among them, trees, due to their size, convert the carbon dioxide to oxygen, which is what we need (deforestation is very bad). Too much carbon dioxide is what causes the "greenhouse effect". I know, I know, the world is sooo big that this really can't be that harmful. You're right, it isn't irreversibly harmful yet, and won't be unless we don't allow for the reestablishment of the oxygen-carbon dioxide equilibrium to some degree. And believe me, because so many people in the federal bureaucracy and in business have so much to lose by changing things to give the earth a chance...it probably won't happen...which of course spells trouble for us. Do your part by being energy efficient.

One other aspect of carbon chemistry that is very interesting and important is that of *Carbon-14 dating*. Carbon-14 is formed in our upper atmosphere by the bombardment of *nitrogen-14* with cosmic rays from the sun. Carbon-14 is radioactive. Although not a lot is formed, enough is made to incorporate a relatively constant amount into all living plants. When the plant dies, no more is added. Since C-14 has a half-life of 5,730 years specimens can be assayed for the amount of C-14 they have left in them and a pretty accurate picture of their true age can be determined.

Hydrogen

If carbon is the big daddy of all elements as they pertain to life on earth, then hydrogen comes in a close second and is actually the big daddy of the universe. The reason for this is simply that hydrogen accounts for some 88% of the atoms in the <u>entire</u> universe. Also, if not for hydrogen there would be no solar energy to drive life systems on earth and there would be no water. Hydrogen derives its name from the Greek words *hydro* and *genes*, meaning "water forming". It was said of George Washington that he was first in war, first in peace, and first in the hearts of his countrymen. Among chemists it is said that hydrogen was the first element of creation, is the first element of the periodic table, is the lightest gas, and is special because it forms its own unique bonds to itself and other elements.

Our star, the sun, converts around 600 million tons of hydrogen *per second* into helium. At the center of the sun the temperatures exceed 13 million degrees and the density is about 200 kilograms *per liter*. This is what it takes to get protons (hydrogen atoms) to fuse to make helium, and lots of energy is released in the process. This is the process that drives all stars' energy output.

You have probably heard the planet Jupiter referred to as a "gas-giant". You guessed it, a probe in 1995 indicated that its composition was 99.8% hydrogen and helium. Hydrogen on our earth is light enough to actually escape into space but this doesn't occur on Jupiter. The reason for this is that the planet is so massive that the gravitational pull doesn't allow even something as light as a hydrogen atom from slipping away into space.

Water makes up some 65% of the human body and hydrogen comprises 11% of water by weight which leads me to calculate that we are roughly about 7% hydrogen. Hydrogen is a component of DNA and in fact the reason why DNA assumes and maintains a double helical (twisted ladder) structure is mainly because of hydrogen bonding and its hydrogen content.

Oddly, water can actually be fatal if consumed in excess. This is especially likely if over consumption occurs when dehydrated. You may remember a story a few years back of an otherwise normal teenager that had experimented with the drug "ecstasy" and thinking afterward she had made a mistake reasoned that the best thing to do would be to dilute the drug in her body to minimize its affect.

She did this by drinking much more water than she should have thereby upsetting the sodium/potassium balance of her body which ended up affecting her brain and heart and she died. If she would've imbibed an excess of Gatorade or something with electrolytes, instead of pure water, she would've lived.

Much of the role of hydrogen on earth, especially as it pertains to life, concerns its link to water and acids. It's not too hard to figure water's importance both systemically and historically but the importance of acids is a bit more subtle. The definition of an acid is generally thought of as compounds that give H+ (a hydrogen ion or proton) to other compounds. As with many other aspect of chemistry an exact and proper definition is not so simple. Chemists designate acids and bases as *Arrhenius* acids and bases, *Bronsted-Lowry* acids and bases, or *Lewis* acids and bases. Luckily these definitions are roughly equivalent.

How readily a compound wants to give a proton (H+) determines how strong an acid it is. The pH scale is used by scientists to measure this availability of hydrogen ions in solution and it goes from 0 to 14. It does have a mathematical basis so hold your breath and let me get this out, $pH = -\log[H+]$, you can open your eyes now. What this means simply is that pH's of about 7 are said to be neutral, pH's above 7 are said to be basic (or alkaline) and pH's below 7 are said to be acidic. The lower the pH the stronger the acid and the more corrosive it is.

Lemon juice, coffee, and sulfuric acid are all acids but only one is used as an acid industrially due to its low pH and eagerness to give up H+ and that is coffee (just wanted to see if you were paying attention...of course its sulfuric acid). Conversely, compounds with a high pH are said to be alkaline or they are said to be strong bases. The higher the pH the greater the concentration of OH- ions (hydroxide ions, pH is all about H+ and OH-, the components of water!) and the more corrosive the compound. Our bodies are very close to neutral and blood in our bodies is only slightly alkaline.

Household ammonia (NH3) that is used to clean windows has a pH of about 12 and Drano, which is sodium hydroxide (NaOH), can have a pH of 13 or so. That's what makes it such a good drain cleaner and also why you mustn't let it touch your skin. Bases like Drano or milder ones like Comet cleanser are good cleaners and if they do touch you skin you will notice a slipperiness to them. Other properties of bases are bitterness to taste, usually no smell (except ammonia) and they react with many fats and oils. Properties of acids are sour to taste, sticky to the touch instead of slippery, the smell frequently will burn your nose (especially the stronger they are), and they react with metals to form hydrogen gas so don't light that match.

Some scientists see hydrogen as the clean, efficient fuel of the future. It would be generated from water and when burned, yield water. Energy given off by burning hydrogen is nearly triple that of conventional fuels like say methane (on a weight equivalent basis). Unfortunately there are some obstacles to overcome. Hydrogen gas needs to be transported under great pressure to be efficient and it is very, very flammable. All you have to do is think back to the Challenger Shuttle tragedy in 1986 to visualize the problem. That hydrogen container that blew contained 1.5 million liters of liquid hydrogen.

The Hydrogen cycle, as one would guess, is tied to the cycle of water. The constant cycling of water to earth as rain and back to the atmosphere through evaporation maintains the various freshwater environs and supplies the vast quantities of water needed for life maintenance on land. The water cycle as previously discussed in the sulfur section, is an important part of temperature regulation through cloud formation and is extremely important since it provides a medium for transport of nutrients through the ecosystem. The added importance of hydrogen due to its role in *every* system on earth should be understood and respected. Its role in acid-base balances in all life systems and the unique properties of hydrogen bonding give it a special place in chemistry, on our planet, and in our universe.

Respect is what it is all about and we as humans through our arrogance don't have enough for our environment. It took millions and millions and millions of years for our earth to establish the equilibrium that was evident back in the early 19th century, before the "age of industrialization". The incredible rapid growth of the human population and its arrogant, misguided need for global dominance and resource consumption is screwing up our planet. People who don't want to hear this because they are making money on the rape of the world say things like "the world is bigger than you think" or "it can take it". They really don't have a clue. They have a vested interest in its demise so they can't see past their bulging wallets.

Every measurable biological and chemical system on earth is in a state of imbalance. Oh, and here's a news flash for you. This mess is like a big rock rolling downhill, it has inertia. It's self-propagating because of lobbying interests of big businesses that don't care and people in high places that are uninformed or misguided. Then there's the added burden of not being able to go back to the way it was unless we leave it alone for millions of years. And it looks like things are going to get worse. The World Bank has predicted nearly a tripling of earth's population by 2050! Some United Nations estimates forecast 9 billion by 2025. This imbalance that we as humans have created is taking a toll. Species that have inhabited the earth for eons are disappearing at an alarming rate! There actually is a worldwide effort by scientists working together to identify and categorize and catalogue species of plant and animal life before they become extinct. I'm not sure you got what I just said. These are species of flora and fauna that haven't even been discovered yet. Scientists understand that deep in the areas where man is developing acreage, say in the Amazon rainforest, the local environ is being so damaged that species' tenuous hold on life is being cleaved before we even know they existed!! It's all because of entrepreneurs responding to economic incentives. I understand that this is capitalism at its finest and is the American way. I'm proud to be an American and am proud of all we've accomplished. That being said, we just have to be smart enough to factor into our appetite for change and development, a strategy for co-existence with our planet and other life forms on it.

Unbridled use of antibiotics in man or what we eat (widespread administering of antibiotics to cattle, pigs, and poultry to maximize profit), disruption of pH levels in our lakes, waterways, and oceans (via sewage dumping, waste misman-agement, and overuse of fertilizers), and our overdependence on fossil fuels (to drive our cars, heat our homes and drive our turbines) is speeding us toward a very sick planet.

Diversity on our planet occurred for a reason and it is this diversity that allows us to maintain our planet's health. Each species in their own way nourishes the world and is in turn nourished by it. When the ecosphere is compromised in some way the symptoms become evident by sudden species *dieoffs, desertification,* or *epidemics* affecting both man and other species.

We've had all these. Worst on the list is epidemics. We now have *MRSA* (methicillin resistant *S. aureus*). *Staphylococcus aureus* is a common type of bacte-ria that resides on all of us. It's not a problem unless you happen to get it in a cut, wound, burned skin area or you are an *immuno-compromised* individual. If it overgrows you use antibiotics to combat the infection. If the bacteria happens to have acquired the ability to "resist" or be relatively unaffected by the antibiotic your using then you use a different one. If you keep doing this and none of them work, you used to be able to go with the big-bopper, (expensive) methicillin, a cousin of penicillin. Because of overuse of antibiotics as of 1992, 40% of *staph* in hospitals are resistant. It used to be that if methicillin didn't work then you went with the last bastion, which was vancomycin. Even that is losing its effectiveness.

H. C. Neu wrote some very scary things in "The crises in antibiotic resis-tance", *Science* 257 (1992). I won't belabor this by overwhelming you with any more statistics and grim findings except for this one…pharmaceutical companies aren't even developing anything promising. There really isn't anything of significance in the pipeline. Ironically, there isn't any money in it. Companies rather develop drugs like *viagra, propecia,* and *botox.* Drugs they know will be used or consumed regularly for a long time so they can make lots of money and satisfy shareholders. Antibiotics are expensive to develop and are only taken by people briefly when they get sick. Worse yet, if you develop an antibiotic that is really effective but terribly expensive, the company looks terrible when they don't give it to poor people that are dying.

Suffice it to say that the crisis is coming. There is now mutated *Hanta virus* in our mouse population, *West Nile* virus in our crows and bluejays, *Ebola* virus that we don't even know where it comes from, *Legionella* bacterium causing Legionnaires disease from our air conditioners and water supply, toxin releasing *E. Coli* in our cattle and hamburgers, toxin-releasing autoimmune disease causing *staph* associated with tampon use (toxic shock), HIV, all sorts of Hepatitis, and cytomegalovirus in our hemophiliacs by way of donated blood from IV drug users and promiscuous homosexuals and heterosexuals the world over, quinine resistant *malaria, Lassa* fever, *Dengue* or *Breakbone fever,* and this list goes on and on and on.

These viruses and bacteria causing new outbreaks of disease that we can't immediately control is simply a symptom of a greater problem…an ecosphere out of balance. The acid-base balance of the world, through the mismanagement of the hydrogen cycle, is askew.

Here's a great example for you. Ocean mammals like harbor seals and porpoises died and washed ashore onto beaches from the Gulf of Mexico, along the Eastern U.S. coast to the St. Lawrence Seaway. This was in the early 1990's and there were hundreds and hundreds of them in groups, dead. No one could figure out what it was. Finally they pieced it together. El Nino weather patterns caused an unusual amount of rainfall in the Midwest which led to substantial flooding. The cattle industry's waste was sitting around in flooded areas and it was all deposited into the Mississippi which then deposited a good share of it into the Gulf of Mexico. Along with all this waste nitrogen was lots and lots of industrially generated sulfur-nitrogen fertilizers. PCB's and other chlorinated hydrocarbon containing toxic chemicals also were included. A nice cocktail for our oceans, huh? All this high nitrogen stuff caused a large increase in algae populations, especially certain types that released toxins. As this mess circled Florida and moved with the currents it seemed the PCB's lowered the immune-fitness of these sea mammals and they fell prey to the toxins. This all occurred because of an acute localized disruption in the equilibrium. There are literally hundreds and hundreds of examples like this. All symptoms of a world that is sick and getting sicker.

Ecologists Paul and Anne Ehrlich of Stanford University developed what they called the "Rivet Hypothesis" to explain the problem. Picture a huge airplane (our ecosphere) flying through the air held together by rivets (the species of plants and animals). As species disappear or rivets fall away from the plane, a point is reached that the entire plane falls apart, or our planet becomes nearly uninhabitable.

What can we do? Unfortunately it would take a worldwide effort by all inhabitants and until things get much worse, no one is going to listen. Individually, little things like completing your antibiotic for the entire course of medication, cleaning wounds immediately after occurrence, being well nourished, washing hands often, and staying in tune with local healthcare workers, is a start. Lobby lawmakers and insist they quit allowing the administration of antibiotics to animals for profit's sake. Vote for anything that controls world population (within reason of course), limits harm to the ecosphere, and otherwise runs counter to the balance on our planet. This is not much but anything will slow our demise. After all it really comes down to the environmental impact of each individual that collectively disturbs. Now cross your fingers and hold on…we're going to see soon what the world can take…let's hope its ability to withstand disruption is greater than most scientists think it is.

Nitrogen

Nitrogen is the seventh element of the periodic table, has seven protons and seven electrons and seven neutrons. If your lucky number is seven, I guess it's your element. It has five out of eight electrons in its outer shell and therefore loves to bond and react with itself and lots of other elements. It occupies an integral niche in the biochemistry of all living things due mainly to its place in DNA and amino acids. Incredibly, it is all around us in its inert, diatomic form, N2, which comprises about 78% of the air we breathe. A few percentage points more and we would asphyxiate.

Its name is derived from the Greek words *nitron* and *genes*, meaning "nitre forming". Nitre was the old name for potassium nitrate (KNO3) which was more commonly known as *saltpetre*. Saltpetre is an important constituent of gunpowder. *Nitrogycerin*, *TNT*, and other explosives all have nitrogen as a part of them. This is due to nitrogen's eagerness to break bonds with other elements and to revert violently to nitrogen gas with the release of large amounts of heat.

Amino acids are the cornerstone of life. In fact you can practically define life as anything that passes on DNA from one generation to the next. Now I was just talking about amino acids and then I talked about DNA as if they were the same…why? They are far from the same at the molecular level. The catch here is the flow of information in life. Information in the genetic code of DNA flows from DNA, to RNA, to protein. Proteins are made up of amino acids. There are 20 common and several rare amino acids that make all proteins in all living things. There are over 150 other amino acids known to occur biologically in free or combined form but never in proteins. DNA in our cells, amazingly, can replicate itself into identical daughter molecules, transcribe itself into messenger RNA strands (that then flow out of the nucleus of the cell to the cytoplasm to waiting *ribosomes*), and be translated to proteins. The flow of information in life then, most of the time, is DNA to RNA to Protein via *replication*, *transcription*, and *translation* (exceptions, though minor, involve certain viruses that have their genetic material in the form of RNA and therefore add a reverse transcription step to get to DNA and then it's the same). This is all very complex, I admit it. So

I will stop and say that simply nitrogen is very important in all this…especially amino acids.

How does its role in amino acids and protein affect us? You know the basic difference between carbohydrates, like the simple and complex sugars, and proteins, that are full of amino acids, is just the nitrogen. You take away the nitrogen from an amino acid like *alanine* or *histidine*, modify it slightly and you know what you have? You have carbohydrate fuel for your cells. Anytime you eat too much protein, beyond what your body needs, it simply cuts the nitrogen off the molecule (*deamination*) and pretty much treats it as a carbohydrate then. The excess nitrogen is then eliminated from your body via the urine. This is why the Adkins diet is really just like all the other fad diets. You still derive 4 kcal/gram of protein and 4 kcal/gram of carbohydrate. Please don't fall for all the trendy diets…just remember this, be active, take in fewer calories than you consume and you will lose weight.

Nitrogen in the form of ammonium, usually combined with sulfur, as ammonium sulfate, is known by all as fertilizer. The only sources of fertilizer before the industrial age and mass over-production of fertilizers was animal waste, nitrogen fixation by bacteria, and conversion of N2 in the atmosphere to organic (useable) nitrogen via lightning strikes.

Remember that the atmosphere is 78% nitrogen in its diatomic form. This means that although nitrogen has only 5 electrons out of 8 in its outer shell, when it combines with itself it's "happy and content" and no longer reactive. In other words it's inert. It takes special occurrences like the energy from a lightning strike or enzymatic conversion by bacteria to change diatomic nitrogen to a form that is useful.

Nitrogen fixation, as mentioned previously, is a big player in this area. Before mass production of fertilizers, farmers would grow crops one year that enriched soil with nitrogen, harvest just a small proportion of them and till the rest back into the soil. The next year they would grow their money crop in the enriched soil and obtain a much higher yield per acre by doing this. By sectioning off their total farmland into, say, four areas and rotating through the sections yearly, they would optimally manage their land and would also harmonize with their environment.

Nitrogen fixing bacteria living *symbiotically* in the roots of certain crops (clover, legumes such as beans, soy beans, and alfalfa) make this possible. Bacteria that can turn this trick have enzymes called *nitrogenases*.

These nitrogenases are comprised of two *polypeptide* chains, one larger than the other, with atoms of iron and molybdenum added. The smaller one contains four iron atoms while the larger one contains twelve. The larger also has two molybdenum atoms. By using the term polypeptide all I meant was that, like a string of pearls, amino acids form a chain. They don't stay in a straight laid out shape but instead achieve a three-dimensional configuration like a long spaghetti noodle piled on your plate. The elements iron and molybdenum find the <u>same</u> place to sit in every nitrogenase molecule that is made and every one achieves the <u>same</u> conformation or 3-D shape. This is the same with all proteins whether they're *immunoglobulins* (antibodies) looking for invading viruses in your blood, *insulin*, a hormone doing its job controlling sugar levels in your body, or *chlorophyll* in plants, capturing light and converting carbon dioxide to oxygen and thereby allowing plants to grow.

The way they do this is really no mystery. The 20 amino acids all are the same in one part…the part with the nitrogen and carboxyl ends, and they are all different from one another on the remainder of each molecule. This remaining part can be acidic, basic, or neutral. This is what causes a polypeptide to fold the same way every time. It's because the environment in which they are made has the same salt concentration and pH every time and so as the chain is made (translated from mRNA) on the ribosome it folds and shapes the protein or enzyme depending on each component's interaction with the surrounding environment. It then has its necessary 3-D shape with any trace elements in it and is ready for work. It's not as complex as it seems but since there are so many things being made and so much is going on at any one time in any given cell or organism it appears unbelievably complex. Proteins do so many jobs in life that libraries can be filled with the information and research on them. That's all I'm going to say about them now even though they are among my favorite topics.

The Nitrogen cycle is, as expected, very critical to life on our planet. It is different than the Carbon cycle in that every turn of carbon in the carbon cycle involves the return of CO2 to the atmosphere. This isn't so for nitrogen. Nitrogen can cycle repeatedly from plants to decomposers to nitrifying bacteria to plants again without having to return to the atmosphere (gaseous N2). Although nitrogen doesn't need to return to the atmosphere, a steady amount always does, which ensures the amount of diatomic nitrogen in the atmosphere stays relatively constant. This occurs because there are lots of certain types of bacteria that do the reverse of nitrogen-fixing, called *denitrification*. These bacteria convert ammonia, nitrate, or nitrite (NH3, NO3-, and NO2-, respectively) back to N2 and release it into the atmosphere. This is part of a cycle that also includes amino acid reservoirs in living and recently dead organisms. The major pools then are nitrates, nitrites, ammonia and urea, and amino acids and the large diatomic nitrogen

(N2) pool. The main players causing change to the reservoirs are plants and animals, nitrogen fixing bacteria and nitrifying bacteria (which change the oxidative states of available nitrogen) and of course the denitrifying bacteria which revert nitrogen back to N2.

The world achieved a nice homeostasis or equilibrium, after the last ice age, as far as nitrogen is concerned. When hunter-gatherers turned to an agrarian lifestyle for the first time, about 10,000 years ago, in the "fertile crescent" (from Israel to the foothills of the mountains of Iran), they knew nothing of fertilizers and nitrogen. Over time they came to realize that their domesticated goats and sheep waste products made their crops grow better. Other farmers followed suit in other areas and there really wasn't much of a problem until the Industrial age, when human populations started to really climb and industrial production of fertilizers occurred.

Now, as evidenced by the higher and higher incidences of ecological upsets touched on briefly in the Hydrogen section, it is becoming more and more apparent that we need to temper our zest for bounty with some common sense. Unfortunately it's a little too late. Our population is too large on our planet and the only thing that can happen to the population is natural reductions through natural disasters like famine and disease. It's unpredictable when it will happen or where, but it will.

Phosphorous

The name is derived from the Greek word *phosphoros*, meaning "bringer of light". This was also the name the Greeks gave to the planet Venus, the Morning Star, which of course they did not understand was a planet.

Phosphorous is really an amazing and versatile element. It has an atomic weight of about 31, its atomic number is 15 (it has then 15 protons), and it is very reactive. There are two isotopes, P-31 which is the naturally occurring form and P-32 which is produced synthetically for research and has a half life of only 14 days. Industrially, it can be manufactured in three relatively pure forms, white, red, and black. The color differences are due to variations in its chemical structure, white being by far the most common. White phosphorous is extremely poisonous and will spontaneously combust when exposed to air.

It was first discovered or isolated by Hennig Brandt, in Hamburg, Germany, in 1669. He did this by evaporating urine and heating the residue until it was red hot. He then collected the gas that was given off and condensed it in water, thereby producing phosphorous powder. Why he was heating up urine no one has ever said, and I don't really want to know…I suppose he was just very bored one day. This was the way people made phosphorous for years, though, until it was discovered that bone was full of it. Later, by the end of the 19th century, phosphorous was being extracted from mineral phosphates simply by heating it with coke (not coca cola or cocaine but rather a derivative of coal or pitch) in an electric furnace.

When phosphorous was first discovered and for many years after, it was considered a valuable medicine if given in small quantities. Quacks claimed it would help anything from depression and epilepsy to tuberculosis and cholera. It actually had no therapeutic properties for what it was being used for but that didn't stop people from trying to make money off it. This was a common theme throughout history, though…when something new was discovered, people would market it at once and ask questions later. I guess that's why we have the FDA.

The property that phosphorous has of igniting in the presence of air led to the match and to the great demand for phosphorous in the latter part of the 19th century. In the 20th century this incendiary property of this element was used in warfare to make tracer bullets, phosphorous artillery shells, and the incendiary bomb. In fact, in response to the merciless bombing of major cities in Britain by Nazi Germany during World War II, in an effort to force Germany to capitulate later in the war, firebombing was done by the Allies. In one week in July of 1943, Allied bombers dropped over 2,000 tons of phosphorous incendiary bombs on Hamburg, laying the city to near total ruin. This strategy was used against other cities in the European theatre and in the Pacific theatre as well, with devastating effect.

An additional use, in the context of war, that phosphorous has, is as a nerve gas. I have seen the affect that organophosphate nerve agents have on rabbits (in graduate school a movie made in the 60's was shown to the class in Pharmacology) and it wasn't pretty. These chemicals fall under the category *acetylcholinesterase inhibitors* and work essentially as follows. Acetylcholine is released by nerves when either voluntary or involuntary muscles are employed to do work (ie. any movement or whatever). These molecules, neurotransmitters as they are called, are released from synaptic knobs of nerve endings into the synaptic cleft where they diffuse to the post-synaptic membrane of muscles, causing depolarization of the membrane. This depolarization allows contraction or movement. The acetylcholine, after it diffuses and causes depolarization, is broken down by an enzyme called acetylcholinester*ase* (the *-ase* on the end of nearly anything in biochemistry indicates that the entity is an enzyme that does its job by cutting something apart or cleaving a section of it away to render it inactive).

So, cutting to the ch*ase*, what happens if you are exposed to an acetylcholinesterase inhibitor, like an organophosphate nerve agent, is that after movement of any muscle you would be unable to relax the muscle because the enzyme that normally would stop contraction from continuing would not be available. Hence, you would convulse one muscle system at a time as you react to what is happening until while convulsing completely and unable to acquire a breath and replete with acidosis from the overworking of your muscles, you would die a terrible death of asphyxiation. So avoid these at all costs. Amazingly, up until ten or so years ago, many insecticides employed derivatives of this. After countless accidents and lawsuits the pesticide industry got smart and changed to something less toxic to be available over the counter.

The most amazing and biologically important part of being a phosphorous atom, though, is its role in energy storage and consumption in living organisms. True, it is part of DNA albeit a small part, and yes, as calcium phosphate it comprises nearly 8% of bone in humans, and certainly as a component of membranes in the form of phospholipids it is important, but as the "energy currency" it is an integral part of every cell in your body and in fact every cell in every Eukaryotic (more evolved than bacteria) organism.

As it turns out phosphate bonds are formed to store energy in a readily usable form usually coupled with adenosine.

Adenine, one of the four bases that comprise DNA (along with guanine, cytosine, and thymine), when coupled with a ribose sugar, is adenosine. I mention this here only to illustrate how the body conserves everything, wastes nothing, and has as many components interchangeable as is thermodynamically possible.

The most abundant form of this readily available energy source is *ATP* or *adenosine triphosphate*. This adenosine is coupled with three phosphate atoms and ten, yes ten oxygen atoms. There is also *ADP* or *adenosine diphosphate* and as one might guess *AMP* or *adenosine monophosphate*. As phosphate groups are joined to AMP to make ADP and ultimately ATP, more energy is stored in the bonds and is readily available to cells that need energy and that possess the enzymes (ATP*ase* because they cleave) necessary to release this energy. By this mechanism (the storage of energy via the formation of ATP from the energy-yielding oxidation of fuel molecules like sugars and fats) **chemical work** as in the biosynthesis or construction of anything the cells need to sustain themselves and grow, **osmotic work** as in the maintenance of chemical gradients through active transport, and **mechanical work** as in muscular contraction, are possible.

Now I do not mean to say here that you eat food, and from this ATP is made and stored in your body for use when you need it. That's kind of true but so much more is involved that this is just a simplification to illustrate the importance of phosphorous. More accurately what happens when food is eaten is this: the food is broken up by your teeth during mastication (chewing), and is mixed with saliva. This saliva lubricates the food allowing easier passage into the esophagus and also contains amy*lase*, which starts the breakdown process of starches. The food *bolus* is then swallowed by a complex interplay of muscular movement and the food ends up in the stomach. The stomach, through muscular contraction, continues the work of the teeth by churning the food, mixing it and breaking up the larger pieces. Cells in the stomach lining release mucous, other cells release enzymes, while still others release acids. All this lends toward the further simplification of the food by breaking it down further and further. Some absorption occurs in the stomach but not much. Most absorption occurs in the small intestine which is where the soupy mixture, that is created by the stomach, ends up.

When absorbed from the intestines into the blood, the food is in its component form like glucose, or fatty acid chains, or amino acids. This is where it gets really complex because the parasympathetic nervous system (part of the autonomic nervous system and a complete topic on its own) gets even more involved than it already was, and the endocrine system (the hormonal or humoral response) does likewise. Hormones and enzymes are released that allow and direct the storage and elimination of all things either good or bad. So whether insulin is released from the pancreas or alcohol reduc*tase* from the liver, or any of the countless other enzymes or hormones needed, the end result is that sugar in the form of glycogen (lots of glucose) is stored in the muscles and liver, fats are stored in the liver and adipose tissue, bad things are separated out and eliminated, and everything needed to grow and thrive is sent to where it needs to be sent.

In response to a demand for energy because you are running a race or something, glucose is released from the liver, or mobilized in the muscles, and through your respiration of oxygen, glucose is broken down or "burned" through *oxidative phosphorylation* and our good friend ATP is formed from pools of AMP and ADP and hence work is done.

Reservoirs of energy in the form of phosphate bonds are not exclusive to AMP, ADP, and ATP. That would be way too easy. There are also sinks associated with *guanosine* (a ribosed guanine from DNA), *cytidine* (a ribosed form of cytosine from DNA), *creatine* (phosphocreatine is abundant in muscles and nerve tissue), *phosphoenolpyruvate* (a derivative of pyruvic acid which is part of the standard energy pathway called Glycolysis), and *3-phosphoglyceroyl phosphate*, to name the most important.

Because of phosphate's importance to every cell, you need to make sure you get enough of it. Please don't worry about this though because just about everything has phosphates in it. The foods richest in phosphates are tuna, salmon, sardines, liver, turkey, chicken, eggs, and cheese.

The Phosphorous cycle is unusual and different from the other Chnops cycles because there is no movement of phosphorous into the atmosphere. It cycles from the soil to rivers to oceans and then to the bottom sediment where it accumulates until it is moved by geological uplift! This process takes millions and millions of years. It was once thought to be a major player in the pollution of our planet since it was for years a component of detergents, only to be eliminated in the phosphate-free detergent era that followed. Research since showed that the problem was not the excess phosphates that were released into the environment but rather all the other pollutants such as heavy metals or pesticides which ended up killing the zooplankton which is what usually consumed the excess phosphates.

Phosphorous is the 11th most abundant element on earth and we have nearly two pounds of it in our bodies.

Oxygen

After hydrogen and helium, oxygen is the third most abundant element in the universe. This is due to the consumption cascade or fuel cycle in stars. Hydrogen is burned through a fusion reaction to helium which is burned or fused to carbon which is burned to oxygen.

Higher levels of oxygen, on any planet in the universe, are taken as a sign that there is life on that planet. This is due to the nature of photosynthesis. There are appreciable levels of oxygen on only two other planets in our solar system, Mars and Venus. The amount of oxygen in the atmosphere of Mars is a little over 100 times less than earth's. In Venus there is even less. These are not considered high levels and are due to ultraviolet radiation acting on other molecules. In order to support life an atmospheric concentration somewhere between 16 and 25% is necessary. Our atmosphere has about 21%.

The word oxygen is derived from the Greek words *oxy* and *genes* meaning "acid-forming". Oxygen makes up about 60% of the weight of the human body. It has 8 protons, 8 neutrons generally, and 8 electrons. There are 3 naturally occurring isotopes: O16, O17, and O18. This is due to variation in the number of neutrons. O16 is by far the most abundant. Oxygen belongs to the same group (16) of the periodic table as sulfur and so it shares properties with it. It has 6 electrons in its outer valence shell and therefore loves to react. In fact, it will form oxides with all other elements except helium, neon, argon, and krypton.

Oxygen is unusual as an element because it is so essential to our survival. This is due to the minute by minute demand our bodies have for oxygen, or O2. Our respiratory system, and those of all oxygen-based life forms on earth, has evolved clever mechanisms to ensure that we get what we need to respire adequately.

For humans and other mammals our lungs are the obvious center of this. Through contraction and relaxation of the muscle referred to as the diaphragm, located at the base of the thorax or chest cavity, oxygen is drawn into the lungs and waste expelled from the lungs, much like the action of a bellows. Expansion of the rib cage also aids in this activity.

The lungs are much more sophisticated than mere sacs, though. Air goes through our mouth or nose, through our pharynx, and down into our trachea, or wind-pipe. From there it follows the branching into bronchi leading to two separate lungs. The bronchi divide into bronchioles which further subdivide and subdivide until very small sacs called alveoli are encountered. There are an unbelievable number of these yielding a total surface area where oxygen (O2) exchange can takes place of up to and exceeding *1000 square feet*!

It is here where the fun begins. Each little sac or alveoli is made up of many cells yet is only a cell in thickness and there is a dense bed of blood vessels called capillaries around this tiny balloon-like structure. Oxygen entering the alveoli actually dissolves in the liquid on the surface of the alveoli thus facilitating the uptake into the blood. This is done by simple diffusion across the alveolar membrane into the capillary. So the oxygen is going from an area of higher concentration (alveoli) to lower concentration (capillary) and conversely carbon dioxide goes in the reverse direction by simple diffusion also so we can expel it as the waste product of respiration.

This is how oxygen gets into our blood but once it's there it is even more impressively managed. Transport of oxygen in the blood is accomplished by the use of a protein called *hemoglobin*. Hemoglobin, for the most part, is located in Red Blood Cells and has four subunits, two alpha subunits which are identical and two beta subunits which are identical. Each has an iron (*heme*) atom fastened in a particular location and it is this *heme* entity that allows binding of oxygen. The complexity only starts here though, because hemoglobin can occur in what is called a tense (T) state or in a relaxed (R) state. The T state favors unloading of oxygen and the R state favors loading and keeping the oxygen. In the lungs the pH or acid-base balance is most conducive to loading yet the hemoglobin, returning from peripheral tissues, is in the tense state. This doesn't mean oxygen can't load but only means the concentration of oxygen needs to be higher which of course it is. When hemoglobin picks up its first of four oxygen molecules, incredibly, it undergoes a conformational change toward the R state. This means its actual 3-dimensional structure changes (from T state to R state). This is a phenomenon called *Co-operative binding* and a complex mathematical formula involving concentrations of oxygen and logarithmic functions can be used to predict loading and unloading by respiratory physiologists. Each successive oxygen molecule that is loaded makes loading of the next easier until all four are loaded and off to the peripheral tissues it goes to deliver its load.

Once in the peripheral tissues whether muscles, spleen, skin, or whatever, it unloads the oxygen in a process which is pretty much the reverse of what happened in the lungs. What makes it happen, though has mostly to do with the pH (or hydrogen ion concentration or acid base balance whatever you want to call it) of the peripheral compartment. You see, one thing that cells do is use oxygen and thusly they give off carbon dioxide (CO_2) as a waste product. The peripheral compartment is abundant with water which when combined with the CO_2 yields *carbonic acid*. This means that the area where the oxygen is delivered to is *more acidic* than where it was picked up (the lungs). This causes the four part hemoglobin *tetramer* to want to revert back to the T state which of course means all oxygens off the bus. As each oxygen is freed, the likelihood of the next being freed goes up and via conformational change it reverts back to the T state completely.

If this process is occurring in muscles that are very active and lactic acid is being produced as well as CO_2 then due to the even lower pH the offloading of oxygen and reverting back to the T state occurs even more rapidly.

Part of this clever mechanism is the added bonus of what is called carbamylation. This is where, due to the lower pH, some CO_2 molecules are actually added to select areas of this four part protein and hitch a ride back to the lungs where they are released and exhaled as our exhaust. This is important because it facilitates the usual simple diffusion of CO_2 into blood and simple transport back to the lung as a dissolved blood gas where it is eliminated.

Wait, though, that ain't all. Adding to the complexity or beauty, depending on perspective, is the role of *2,3 bi-phosphoglycerate* (often called *2,3 diphosphoglycerate* by biochemists). This tetramer called hemoglobin, being composed of four parts is sorta like four donuts, laid out, touching. Between all four, in the middle, is a space. This happens to be the area where 2,3 biphosphoglycerate, or BPG for short, likes to reside. It "likes" it here only due to the fact that it is a highly *negatively* charged molecule and in the internal aspect of the four part hemoglobin molecule, *positively* charged residues of each subunit, reside. In fact, in deoxyhemoglobin (hemoglobin without O2, or the T state), because of the conformational change, this internal space is larger, and BPG is readily bound. BPG also is an intermediate of the breakdown of sugar, so when, say, a muscle is working hard, you can expect to find lots of BPG in the area. Conversely, you don't find much in the lung. The bottom line in BPG's role is the shift in the equilibrium between the T state and the R state causing more oxygen to be dumped in the peripheral tissues at a lower oxygen level.

People living at high altitudes have a much higher BPG level than people at sea level. Also, people that chronically smoke cigarettes adapt by increasing BPG levels. "Blood doping" by athletes that run long distance races or ride "Tour de France" type races, often, in the past and presently, involve BPG because it is natural and nearly untraceable.

An additional fact that must be understood is the relationship between hemoglobin and Red Blood Cells (RBC's, also called erythrocytes). Although the hemoglobin tetramer is just as efficient carrying oxygen when dissolved free in blood as when it is in a red blood cell, there is a decided adaptive advantage for most hemoglobin to be in RBC's. It's simply that more hemoglobin can be carried per unit volume of blood when they are packaged in RBCs than if all hemoglobin was just dissolved in blood. This really increases our oxygen carrying capacity. If there were no RBCs and all our hemoglobin instead was just dissolved in our blood, there would be such a high protein concentration in our blood that it would be as thick as syrup...and hence we would need a heart 50 to 100 times larger and more powerful to force the blood peripherally.

One more thing, briefly about the hemoglobin-oxygen delivery system that I must mention, because it is so amazing, is the type of hemoglobin babies use *in utero*. Fetal hemoglobin is different from regular hemoglobin in that it has the same two alpha subunits but instead of beta subunits it has subunits designated as gamma subunits. Since babies *in utero* are completely dependent on their mother's blood for oxygen, this fetal hemoglobin has evolved to have an even higher affinity for oxygen than regular hemoglobin and thus the baby can always extract the O2 it needs from the mother's blood. After birth the gene responsible for making the gamma subunit shuts down and the gene that makes the beta subunit turns on and the situation is normal.

I would like to go onto the topic of Sickle Cell anemia (a disease caused by a mutation in the beta subunit which causes shape changes in RBCs, makes malaria less of a problem but can cause early death) and then discuss myoglobin (a form of hemoglobin in muscle responsible for the efficiency of O2 delivery to individual muscle cells) but I must get back to the main topic, so I won't.

So we know basically how higher animals "breathe" but what about fish and insects? Most aquatic animals have gills, and although there are variations, they do the same thing which is to extract O2 out of the water. The tricky part of the whole thing is that oxygen has a low solubility in water. In air we breathe, as mentioned before, the concentration is about 21%. In water, due to oxygen's low solubility (it doesn't dissolve well), the concentration is 0.5% or only *one half of one percent*!! (this is in salt water, it is slightly higher usually in fresh water). To make matters more difficult the diffusion of oxygen in water is many thousands of times slower than in air! So how the heck do they do it? They compensate by having an apparatus (the gill) that exposes a large surface area to the water <u>and</u> as anyone that has watched a fish before can tell you, they keep the water moving over the gills constantly. They do this by opening and closing their mouths and gill slits.

O2 diffuses across the 1 or 2 cells separating water from blood and then it's up to fish hemoglobin to get the O2 where it's needed. An interesting difference between fish and mammals though is that blood in mammals first returns to the lung for oxygenation, returns to the heart, and then is pushed out the left ventricle of the heart to the systemic circulation which is where the peripheral tissues are. In fish, the oxygenated blood goes directly to the peripheral tissues and muscles and then returns to the heart (fish have two-chambered hearts instead of four like us) and then deoxygenated blood is pumped to the gills again.

Insects, such as grasshoppers, "breathe" not by the use of lungs or even with the help of a circulatory system, but rather by a system of many small tubes that ramify throughout the body. They rely mostly on passive diffusion although there are many exceptions to this where muscular contraction facilitates to a degree the respiratory process.

Acquiring oxygen this way has been a major factor in the limitation of insects to only a certain size and no greater. It also means that the warmer the climate the larger insects will be. This is due to the effect of temperature on chemical processes (i.e. the warmer the temperature, the greater the rate of diffusion of oxygen). Also, it is the main reason why bugs disappear when the weather turns cooler.

Insects generally have a primitive heart which doesn't resemble our heart. It is basically a long tube, that has some muscular activity, and that runs along their dorsum. "Blood" is pushed toward the head and as it migrates around cells nourishing them with fuel and nutrients, it also accumulates wastes. This liquid migrates back through the thorax and into the abdomen where the wastes are processed and eliminated and the liquid is enriched with fuel and nutrients once again and enters the "heart" at the other end and eventually is injected out toward the head again. There are no veins, capillaries, or arteries. Insects are said to have *open* respiratory and circulatory systems unlike our *closed* systems.

Oxygen, as ubiquitous as it is on our planet, was not well understood incredibly until the latter parts of the 18th century. Before this, the theory of "phlogiston" (I'm not kidding) was used to explain combustion, respiration and the change in reactants when oxidized. Phlogiston was thought to be a weightless or near weightless substance. Metals and fire were thought to be phlogiston-rich and earth or dirt was thought to be phlogiston-poor. Chemistry was barely a science at this time until a Frenchman named Antoine Lavoisier, using information discovered by an Englishman named Joseph Priestly and a Swede named Carl

Scheele, regarding the existence of oxygen, debunked the Phlogiston theory and introduced a different way of looking at chemistry. Lavoisier coined the term *oxygene* and is known as the Father of Modern Chemistry. He perfected the balance for accurate measurements and perfected the calorimeter to measure heat loss or gain during chemical reactions. He was also the first to show that oxygen was consumed during respiration and carbon dioxide was given off as a waste product. He was truly a brilliant mind but unfortunately he became involved in the French Revolution and was put to death in 1794, to be buried in a common grave.

Joseph Priestly (see photosynthesis in Carbon section), the co-discoverer of oxygen, also discovered another oxygen containing gas of interest, called *nitrous oxide*. This gas, now affectionately referred to as "laughing gas", is encountered mostly at the dental office but is also used as a fuel in race cars. A close cousin, *nitric oxide*, oddly enough is important in male erectile tissue vascular dilatation and it is in this system that Viagra and related products work. Go figure.

Nitrous oxide though has quite an amazing story of its discovery and early usage. Although discovered in 1772, it remained obscure for 29 years until another Englishman, Humphry Davy thought up a medical use for it. Thomas Beddoes, a prominent physician in Bristol, England, initiated planning for one of the first philanthropically funded medical clinics called The Pneumatic Institution. He selected a young, poor chemist (Davy) to head the Institution and direct its efforts. The realm of respiratory therapy was the focus. There were lots of lung ailments in those days, chief among them was tuberculosis. Davy selected nitrous oxide as the first gas to explore and although it was thought to be poisonous he decided to try it on himself. After thorough study he shared it with his friends who all liked it immeasurably. In 1800 he compiled a 400 page work on the gas and suggested as a part of it that the gas "may probably be used with advantage during surgical operations".

You see, at that time, because of the limited knowledge of infections and bacteria, people were often having limbs, tumors, and teeth removed without anesthesia. If you got an infection in your leg and it was bad enough, they just sawed it off and cauterized it with heat or silver nitrate or both. When they did this they chained you down or held you down and cut it right off. The term "sawbone" was synonymous with physician and a good one could get through a limb with two maybe three swipes with a saw, tops. Many people chose to die from their infection or problem rather than go through the horror of surgery.

Some tried exsanguination known as "bleeding to *deliquium animi*". This was a dicey procedure where a cut was made in a large vessel of the patient's and blood was drained until the patient passed out. The infected appendage or problem area was then removed and the patient was supported with nourishment and rest until hopefully they recovered. Needless to say this was unpopular but was advanced in the 1820's and performed often by an English physician named James Waldrop who was no less than the Prince Regent Physician.

Another approach was advanced by an American physician living in London around the same time. He was apparently inspired by the story of a colleague who observed an Irishman that became so drunk that he fell unconscious on the ground in an area frequented by pigs. During his sleep, part of his face was eaten away by a large hog without rousing the drunkard. Alcohol was tried often but was found to be inadequate. Opium was used also but people kept dying from overdosage.

Now most people know that a little nitrous oxide can make you feel light headed but did not know that people could be put completely out, have a procedure done on them and then be aroused to normalcy afterward without memory of the event. The reason for this is simple...nitrous oxide, because of its close resemblance to O2 or oxygen, can be picked up in the lungs by hemoglobin instead of the intended oxygen. In fact, as a quirk of nature nitrous oxide is preferred by hemoglobin. Yes, if equal amounts of oxygen and nitrous oxide are present in the lung, much more nitrous will be grabbed and carried by hemoglobin to peripheral tissues, including the brain. The brain needs lots of oxygen constantly. When it is given mostly nitrous and doesn't get enough oxygen, it turns the lights out. This means the person passes out. In dental offices many safety features are incorporated into the oxygen-nitrous oxide delivery system to make sure you can't get too much, just enough to make you light-headed. In the past though people completely passed out and many died from over administration. As an aside carbon monoxide is so deadly because hemoglobin binds it much, much more readily than O2 and then it has difficulty releasing it. The nice thing about nitrous oxide is that it grabs it but it also releases it predictably, making it relatively safe.

So this Humphry Davy guy studies this new gas, publishes a long report on it and concludes that it doesn't help respiratory problems much, it is a lot of fun, and oh, it could possibly be used to alleviate the horrors of surgery. This as I mentioned, was 1800. It was pretty much dropped as a therapeutic agent for breathing problems soon after but enjoyed an incredible surge in popularity as a "fun gas". No one, ironically, until 40 years or so later, ever thought to use it in surgery!! Like Tom Cruise demands of Jack Nicholson in *A Few Good Men*, "can you explain that!?"

You see, people in the medical field had dealt with the problem of pain in surgery for so long it was accepted that that was the way it was ordained to be. Famous French physician and teacher, Alfred Valpaeu, at a leading Paris Medical school, said in 1839: "Knife and pain in surgery, are two words which will never present themselves the one without the other…and it is necessary for us surgeons to admit their association".

So from the early years of the 19th century to roughly forty years later the sole usefulness of nitrous oxide was to make people feel different. It was a diversion from life. Promoters announced a party in a certain hall in a certain town at a certain time and you could come and hear about it for ten cents or so and experience it. It became all the rage in Europe and America. In 1833 an Albany, New York newspaper published a review of a successful demonstration of "Laughing Gas" by a Dr. Coult of London. In reality it was actually Samuel Colt, the inventor of the revolving chamber Colt handgun! He had invented his improvement for guns a year or so before and did this "gas passing" to raise money to apply for a patent and get his business going. He actually started by selling hits of the gas on the street corner of some towns for a nickel! Others did this also. It was quite popular.

Around the same time *Ether*, another oxygen containing compound, was discovered. Neither gas was illegal and did the same thing, basically, just in a different way. It was at one of these demonstration parties in Hartford, Connecticut, that a dentist named Horace Wells, attending out of curiosity, with his wife Elizabeth, observed a friend sampling nitrous oxide. He was in line to try it himself when he observed his friend partake and then dance crazily around, inadvertently colliding a chair with his knee, which immediately started bleeding. When the friend recovered his senses he appeared to have no awareness of his injury. When asked about his bleeding leg by Dr. Wells, the friend asked what had happened and a light bulb went off in the dentist's brain. He, the very next morning, employed the demonstrater at the hall, Dr. Colton (no relation to Colt) to assist him in the extraction of a tooth and Colton consented to attempt to put a patient under. They found a patient, put him under, pulled the tooth without pain or memory of the event, and the Age of Anesthesia was born. Wells went on to demonstrate this at Harvard Medical successfully, though it was not perceived by attendees as such, but soon to follow were others with ether and from then on people suffered much, much less.

I could and would like to go on and on about oxygen's pervasiveness in our surroundings and history…but I won't. Instead I will discuss the Oxygen cycle on our planet and how it affects us.

The Oxygen cycle involves oxygen and all the different elements that interact with it. As was said earlier, oxygen is the third most abundant element in the universe but it is the most abundant element on earth. Our atmosphere has lots of it available for us to breathe but it hasn't always been this way, though. The oxygen in the earth's atmosphere comes from the photosynthesis of plants. This was briefly discussed in the Carbon section so suffice it to say that one of the end products of photosynthesis is oxygen. Early in our earth's history there was a much greater carbon dioxide level than now and very little oxygen. Photosynthesis, according to the geologic and fossil records unearthed and dated, seems to have begun about 3 billion years ago (the earth is thought to be about 4.5 billion years old, according to isotope dating techniques). The amount of oxygen in the atmosphere, though, did not appreciably rise for another billion years or so. This was because of the abundance of iron in the ferrous II form and its "need" chemically to react with any oxygen in air to form ferrous III. About 2 billion years ago the level of oxygen on our planet reached about 1%. This is when the fossil record indicates that blue-green algae (cyanobacteria) first appeared. Atmospheric oxygen concentration was on its way up slowly when about 500 million years ago the first land plants started to appear, which caused it to rise relatively rapidly to about 20%.

Oddly enough, because of the incredible abundance of O2 in the atmosphere, the burning of over 7 billion tons of fossil fuels each year makes almost no direct impact in the amount of oxygen available to us. Plants, during photosynthesis, and animals, during respiration, involve O2 going in opposite directions. It appears, though, that the average Oxygen atom takes part in this cycle only *every 3000 years* or so! This is not to declare that the world is stable and has nothing to fear from pollution or anything, for the aforementioned topic was on primary oxygen availability. Oxygen has secondary jobs to do, too, and it is here that pollutants cause their greatest harm. Ozone is the most obvious and widely discussed oxygen-pollution topic.

While breathable oxygen is in a diatomic (2 atoms) form, ozone is in a triatomic (yes, three atoms) form. The three oxygen atoms in ozone are joined in a V-shape. There is ozone in the air that we normally breathe but it is in very low levels. This is good because it is not good for us and has a mean tendency to damage our lungs during constant exposure. Unfortunately, during warmer summer months, especially in and around large metropolitan areas, the ozone levels can rise to damaging levels. This is due to automobile exhaust emissions of nitrogen dioxide, mainly, interacting with sunlight. Higher levels of ozone also have a negative effect on plant growth.

Ozone, though, does its best work for us in the stratosphere (the area from 12 to 24 miles above us). It is here that it becomes our guardian.

Our sun, which is actually a small star, is constantly converting hydrogen to helium via the process of fusion. This causes radiation to be released from it and it moves away from the sun concentrically as electromagnetic radiation. Part of this electromagnetic radiation comes to the earth as the sunlight that illuminates our days and drives photosynthetic processes, making it possible to live here on earth. Visible sunlight is just a small part, though, of what is referred to as the *electromagnetic spectrum*. Much like waves of water crashing in to a beach, this radiation has wavelengths. This spectrum is subdivided according to wavelength...the *longer* the wavelength, the *lower* the energy. Visible light involves wavelengths from 760 nanometers (a nanometer is 1 billionth of a meter) to 380 nanometers. Infrared light has longer wavelengths and therefore less energy than visible light. Radio waves longer still. Ultraviolet light has shorter wavelengths than visible light and therefore is higher in energy. X-rays and gamma rays are shorter still and have even higher energy values associated.

Ultraviolet light in wavelengths of less than 240 nanometers (nm), up in the stratosphere, reacts with O_2, splitting it into individual oxygen atoms. These single oxygen atoms don't like being like this (unattached) and latch onto O_2 molecules to form O_3 molecules. This ozone is especially good at absorbing wavelengths of electromagnetic energy of 230–290 nm. This decomposes the ozone back to O_2 and O and this cycle repeats. Without this filter in the stratosphere, high levels of radiation would penetrate to the earth's surface with great harm to living cells. A number of pollutants damage the ozone layer, chief among these are chlorine atoms in the form usually of chlorofluorocarbons (CFC's). You may have heard, recently, that there is a growing hole in our ozone layer. This has incredible long term implications for us all as well as our descendants and will be watched very carefully.

The final word on oxygen belongs to an amazing quirk of the oxygen isotopes. As noted before, there are three different isotopes of oxygen, none are radioactive. They are oxygen-16 which makes up 99.76% of all oxygen, oxygen-17 comprising .04%, and oxygen-18 rounding up the remaining 0.2%. Now because of the two additional neutrons O-18 is about 12% heavier than O-16. It may not appear to be a big deal but the sensitivity of instrumentation these days is unbelievable. Also, since we are dealing with samples with atom counts in the trillion trillion range (this is a 1 followed be 24 zeroes...a huge number), 0.2% becomes a large number. It is in this way that by digging down in the geologic strata, we can take samples of dirt, analyze the relative concentrations of O-16/O-18 and determine temperature fluctuations during past eras. This way we can pinpoint when ice ages were and even if a once-living creatures' body, found in higher latitudes, originated from those latitudes. The reason why this is possible is that because of the 12% weight difference of O-18, O-16 being lighter becomes airborne (evaporates easier) and this difference is enhanced when it is colder. Higher ratios of O-16/O-18 than normal, in the stratified polar ice, yields a very educated guess on when an ice age occurred. Also, if a person was raised closer to the equator, then the amount of O-18 in the enamel of their teeth is greater than in a person of who had lived closer to the poles.

Bacteria

Most people know that bacteria exist. They know that a cut that remains dirty becomes infected due to bacteria. They know that food left unrefrigerated for too long spoils due to bacterial growth. And they know that teeth rot and gums swell due to activity of bacteria in one's mouth. Ask most people about the history of our knowledge of these little microorganisms and you won't get much information. Tell people that it wasn't until relatively recently (since 1800) that hospitals were for the sick and you encounter disbelief. Or tell people that the overwhelming proportion (greater than 99%) of babies *weren't* born in hospitals in the nineteenth century and before and there will be amazement.

The scope of knowledge of the scientific/medical community today regarding these little bugs is impressive but as impressive as it may seem it's unnerving to realize that we know much less than we need to. It's hard to comprehend that after all the research in the field of bacteriology in the last 300 years that we still are discovering new *species* of bacteria. According to the University of Maryand's Rita Colwell, one of the world's foremost authorities on bacteria, we have characterized only 2,000 of over 300,000 species thought to exist. And less than 1% of all ocean bacteria have been characterized!

People for ages have suspected some kind of transmissible element in the disease process and although viruses were a very significant part of this microbiological picture, bacteria were the most common "germ" encountered. Hippocrates, the great Greek father of Medicine and the Hippocratic oath, wrote in 400 BC *On Airs Waters and Places*. In this work he explained how when traveling one should pay attention when arriving in new towns as to how the waters that feed the town are situated, and "whether they be marshy and soft, or hard and running from elevated and rocky situations, and then if saltish and unfit for cooking". He proceeded to talk about how towns were situated for morning or evening sun, whether swamps were nearby and how, if all things were considered, he could tell the townspeople what diseases they could expect and during what seasons. He didn't know why, but experience told him what to expect. The bacteria causing Tuberculosis and Cholera were particularly effective ravagers of mankind 2000 plus years ago.

No presentation on bacteria is complete without mentioning one of the worst scourges on mankind (and the causative bug). The *Bubonic Plague*, or Black Death as it came to be known, decimated the population of Europe in the 1300's. What started in China in 1330 somehow, due to increased trade with the orient, made its way to Europe and killed over 25 million people between 1347 and 1352. The bacteria responsible for this was *Yersinia pestis*. It spread by fleas on rats that stowed away on trade caravans. Unfortunately, just previous to this, in many areas cats were declared to be demonic by the church, due to the appearance of their eyes. For this reason, one of the best methods of rodent control was reduced to non-protective levels, and all hell broke loose. There actually was an epidemic in the 6th century also, and later, in the 17th century it would strike again, though not as widespread. With sanitation and public health advances and vermin control it was eliminated on a large scale. Exceptions occurred in several countries even as late as the early 20th century, though. Most notably India had difficulties with this scourge from 1895 till 1918 when more than 10 million people died from it. As recently as 1994 less than 100 people were killed by an outbreak near New Delhi, India, that caused severe panic in India as well as countries engaging in trade with India.

Another bacterium ravaging humans, only more recently, is *Syphilis*. With gonorrhea and Chlamydia, syphilis rounds out the big three in bacterial *sexually transmitted diseases* (STD's) affecting humans. It was thought for some time that syphilis was brought back with explorers in the 15th century from New World Indians. This is due mostly because 1495 is when syphilis was first described medically as an epidemic (during the Franco-Italian Wars) by European Physicians and scientists and it coincides with Columbus's return from his first voyage (we did enough to the Amerindians, what's one more slap in the face).

The truth is that the evidence says otherwise. Syphilis is caused by the bug *Treponema pallidum,* a *Spirochete* or cork-screw shaped bacterium. The exact same bacterium that causes and caused *Yaws*, a childhood skin disease, dating way back to ancient times in Europe <u>and</u> the Americas. Yet there was no hint of syphilis prior to the 15th century anywhere. Also, syphilis was equally devastating to both Europeans and Amerindians during its early years (late 1400's and 1500's). If it had been in the population before, there would've been at least partial immunity to it in either population. This was a time just after the great European plagues of the 14th century. Population centers were filthy and morality was low. People slept around a lot. Godliness had failed they felt and self-gratification was all that mattered. Interaction of partners and population as a whole reached a critical mass that allowed sexually transmitted microbes to emerge.

It was Edward Hudson that changed everyone's mind on this when he published his work "Treponematosis and Anthropology" in the *Annals of Internal Medicine,* in the 1960's. It checks out and many scientists have taken it further to say that there is no way that it could've evolved in the New World. Too bad they couldn't have settled it back then.

The problem with settling anything regarding bacteria "back then" was the limits of vision and the size of the bugs. Humans can see effectively to about one-tenth of a millimeter (which is 100 micrometers). Bacteria, though, are generally about a micrometer in diameter. So there has to be quite a collection of bacteria in one place, a colony perhaps, and the human eye can detect their presence but knowing that the individual components are live creatures capable of transmission, infection, and disease was not comprehensible.

It really wasn't until the Italian scholar Girolamo Fracastoro wrote the book, *De Contagione*, in 1546, that a rudimentary understanding of person to person transmissibility of the common diseases existed. This was of course only by the people who could read (not many) and who read the book (even fewer) and chose to believe it (almost none). Most people chose to believe the reason for epidemics, pestilence, and disease was through divine providence, a higher power, a supreme being with a magnifying glass burning his subjects. Fracastoro was right on, though, as he wrote of diseases being caused by the transmission of individual seeds or germs, different kinds of which were the specific cause of different diseases.

So it was no wonder that hospitals as we know them today were essentially non-existent prior to the 19th century. The further back in time one went, the stranger things became. The only people that received *in patient* treatment in the 1700's and before were usually people that were sick that had no place to go. In fact, care of the sick by society was difficult and became unwanted. A Bishop of Bath and Wells wrote in 1219, that "no lepers, lunatics, or persons having the falling sickness or other contagious disease, and no pregnant women or sucking infants, and no intolerable persons, even though they be poor and infirm, are to be admitted" to his hospital, and he even went on to say that if admitted by mistake "they are to be expelled as soon as possible".

The role that bacteria played in all of this was of course, a major one. Without an understanding of pathogens, hospitals didn't contribute much to recovery. They contributed more to propagation of epidemics through their worker's ignorance. Due to this, most often, hospitals were run by convents and churches. There were few trained nurses. Doctors were educated with basic knowledge of circulation, diseases, and anatomy. An experienced physician was one who could recognize early signs and symptoms and support a patient's immune system through rest and proper nourishment and also one who could saw through an arm or leg quickly. Prior to the nineteenth century, when ether was discovered for use in anesthesia and infection control methods were started, medicine was as barbaric as it was archaic. Exsanguination or *bleeding* most often using *leeches* was a standard in patient care. Surgery centers in hospitals resembled torture chambers where patients that needed something removed were shackled and restrained during the hacking or sawing and then were lucky if they survived post-surgical infections.

Amazingly you could tell a good sawbone by the surgical garb he wore, for if the surgeon came in to surgery with an overcoat that had putrid remnants of splattered body fluids from past "surgeries", then you knew you had a good one. The physician would use the saw from the previous surgery, days earlier, on the present patient never having sterilized the blade. After the procedure care was taken to clean the handle but the blade once again was ignored.

Nurses, if not Nuns involved with the church, often were prostitutes, especially when it came to war. There usually was a group of women followers of the armies. These ladies would do much more than care for the sick. They also cooked for the men and distributed sexual favors as well as sexually transmitted diseases. Minimal training was needed and very little pay was to be made.

This was why Florence Nightingale was so unusual. She was an educated woman from a well-to-do family that shunned traditional married home life to become, gasp, a nurse!! She became a nurse and served the British Army during the Crimean War no less! Because of her education, though, <u>and</u> (and this is a very big <u>and</u>) her association with the British Secretary of War (they were lifelong friends), she wielded significant power in her quest to redesign hospitals to reduce communicable disease and improve cleanliness. Men in power at the war front consented to her advice on changes only because the Secretary supported her emphatically. The result was amazing. Morbidity and mortality rates fell precipitously and the age of the Nurse was born.

Florence Nightingale had paid attention during her education especially as it pertains to disinfection. Just prior to her formal nursing education a pair of landmark studies was done on *Puerperal Fever*. This fever occurred in women just after birth and as we know today is a septic condition caused by a *streptococcal* bacterial infection (septic meaning bacteria occurring in significant numbers in the blood).

The doctors separately documenting the observations of these studies were Ignaz Semmelweis and Oliver Wendell Holmes, Sr. (Holmes's son would fight in the Civil War, receiving wounds in battle on three separate occasions, would pen the poem *Old Ironsides*, and later serve on the U. S. Supreme Court for 30 years). They noted that the incidence of Puerperal Fever was much higher in groups of women that had their babies delivered by Medical students that had been performing autopsies on diseased bodies just prior to assisting childbirth. Therefore, on May 15, 1847, all medical students were required to wash their hands with chlorinated lime before assisting in deliveries. Semmelweis made his observations in a hospital in Vienna, Austria while Holmes was in Massachusetts. This didn't bring an end to Puerperal fever but it reduced it dramatically and ushered in a new era of cleanliness as it pertains to the patient...a lesson not lost on Ms. Nightingale.

Central to the theme of bacteria and disease regarding human civilization is the role of adequate sanitation. Long ago, as populations grew and became centralized in towns and cities, disease caused by bacteria resulting from squalor became more common. Rotting food and filth became breeding grounds for bacteria, but also rats and mice that carried with them ticks, fleas, and lice carrying with them bacterial pathogens. People thought nothing of urinating or defecating in bodies of water that were also a drinking water resource. Bathing was thought to be dangerous and few Europeans ever washed. People wore wool garments and rarely changed or cleaned their clothing. People shared bedding. Centers for sexual activity were established and flourished.

Inadvertently humans were aiding in the transmission of the microbes. Damming rivers and unwittingly creating stagnant water pools where mosquitoes could breed was particularly useful for the establishment of Yellow fever (a viral disease) and malaria (a parasitical disease). Rome had a population of over one million inhabitants by 5 B.C. and soon after disease changed things dramatically. Pneumonic plague, leprosy, tuberculosis, cholera, typhoid fever, and typhus were all thrown at it along with a barrage of viruses. The result was that no city surpassed the million inhabitant mark again until London did in the 1700's. All because bacteria were too small for the naked eye and technology hadn't advanced to the microscopic level yet. The cause and effect of the lack of sanitation contributing to the plagues was somehow lost because of this.

Then it all changed. A Dutchman, Antony van Leeuwenhoek, apprenticed in a linen-drapers shop in Delft, Holland, was introduced to biconvex (two sides rounded out) lenses for the study of fabric weave. This guy was amazing because by the standards of his day he was not well educated or well placed in society. Yet, despite this and maybe because of this, he was able to usher in a new era of science. Oddly, though, microscopes had been around for some 40 or 50 years.

These were true microscopes using two lenses. And although the microscope was refined and "perfected" by another scientist of Antony's day, Robert Hooke, it was van Leeuwenhoek that, using just a single lens, was able to achieve magnifications of 200 times rather than the 30 to 40X of his cohorts.

He did this by being a superior lens grinder and polisher and by the expert use of lighting. Hooke had published his book in 1665, *Micrographia,* in which he showed elaborate drawings of a flea and plant cells (being the first to coin the term "cell") but it was van Leeuwenhoek, with his superior magnification that shocked the world with his publications in 1678 to the Royal Society of England, of bacteria and *Protists* (one celled plants and animals) labeled *animalcules* isolated from lake water, pond scum, and saliva.

Scientists were in disbelief and Robert Hooke was asked to reproduce Antony's finding and did so thus paving the way for wide acceptance of the discovery. The single lens way of magnification died out, though, as double lens microscopes improved. It seems that the only real reason that van Leeuwenhoek was able to make his discoveries using such a primitive apparatus was his incredible eyesight when studying small things.

Well, you would've thought that this discovery would make light bulbs go off in scientist's brains everywhere. With Fracastoro's ideas on transmissible agents and Leeuwenhoek's discovery of the possible agents of transmission in disease, it doesn't seem difficult in hindsight to extrapolate the disease causing scenario. Yet it wasn't until Semmelweis and Holmes that people started to buy into the idea.

The problem was that people didn't know where these transmissible agents came from. Leeuwenhoek and others postulated that they were in abundance everywhere and when stagnant dirty water was left to stand at room temperature then numbers would increase from a few parental organisms. Conversely, a majority thought that they arose from *spontaneous generation*, arising from nothing but dead organic matter. This controversy retarded advancement in science for a considerable number of years!

In 1718, a French microscopist named Louis Joblot laid the groundwork for the resolution of the debate. He used heat for 15 minutes to kill the bacteria in a flask of hay and water and kept it sealed afterward. Nothing grew. Eureka! But wait a second, or rather a century. Scientists that attempted to duplicate the experiment were not exact on the temperature, length of time to heat the mixture, and how to seal the container. Inconsistent results made the theory of "spontaneous generation" persist until finally (really this time) Lazzaro Spallanzani (1775 and 1776) perfected the technique along with Louis Pasteur (1860–64) who showed what the minimum requirements were to sterilize.

Pasteur also showed that air could be reintroduced into the sterile flask as long as entry was through a dust-stopping "filter". Soon after Ferdinand Cohn (1876) discovered the heat resistant nature of bacteria that had the ability to form spores and Robert Koch (1877) laid the groundwork for easy bacterial growth and study. In 1886 the *chemico-parasitic theory* of oral disease was elucidated which showed that food left in the mouth led to increased populations of bacteria. The bacteria in turn released acids as waste products after eating the food which led to the softening of the teeth (decay) and inflammation of the gums and loss of bone (gingivitis and periodontal disease).

Now after all of this talk I know what you're thinking. Bad, bad bacteria...these babies have got to go. Well, just hold on for a millisecond. First of all, they were here before us (long before us) and secondly, as much as it sounds like they are the root of all evil there are only a small proportion that are pathogenic (disease causing). The rest are commensal, which as discussed in the Greed section means they live in, on, or around us without causing problems.

In fact these commensal organisms often times prevent pathogens from overgrowing and causing problems for us by just being there and competing for a site, say the upper respiratory tract, better than the pathogens. And don't be afraid of them because you have to realize that from the tip of your toes to the top of your head there are more bacteria living inside and on you than there are your own cells that make up you. Incredible, huh!? Even scarier is the fact that in a quarter sized area of your large intestine you could find more bacteria than there are people on the earth right now (that's 6 billion, give or take).

So don't freak out, become one with your environment, grasshopper. Realize that the bacteria in your gut are actually helping you by aiding in digestion and they are a source of vitamin K which is important in blood clotting. Bacteria are generally *saprophytic* which means they consume dead matter. This is very beneficial to us. Certain types also, as discussed in the Nitrogen section, live on the roots of certain bean plants and "fix" nitrogen out of the atmosphere thereby making it available to us in an organic form. Bacteria have a central role in the manufacture of foods such as dairy products and also help to make vinegar to name just a few other ways they're beneficial. So really they are just part of the normal environment and we just have to take steps to prevent them from causing problems.

As stated earlier in this section, humans, due to their misunderstanding of their environment centuries ago, contributed to their misery by doing things like living in squalor which led to disease, suffering, and death. Nowadays we are similarly contributing to the problem but in a different way. By saying this I don't mean to say that the degree to which we are hurting ourselves now is equal to what we did in the past but it could become that way. The difference is a quantitative versus qualitative one. In other words in the past the conditions led to a much larger quantity of bacteria which led to certain bacteria overgrowing and causing epidemics. Presently, due to impressive sanitation techniques, our understanding of the pathogens, and the advancement of science, we have to worry more about the quality of the bug we encounter. And unfortunately we are to blame for this.

We are presently in the seventh decade of the "antibiotic age". Individuals that have lived or have grown up during this time have enjoyed the most disease-free time in recorded history, by far. We no longer have to worry about common bacterial diseases like typhus or scarlet fever or bacterial pneumonia because a simple regimen of antibiotics will take care of it, right? Unfortunately, those days, though not completely gone, are waning. And it's simply because we've misused and over used these once effective bacteria-controlling agents.

For example, in 1941, a dose of 10,000 units of penicillin for four days was enough to easily control a streptococcal respiratory infection (*Strep throat*). By 1992 the same ailment often required 24 million units per day and despite this could still be lethal. Remember puppeteer Jim Henson of Muppet creation fame? In 1990 he succumbed to a particularly strong type of Strep that was not only resistant to penicillin but happened to make a killer toxin also. More recently (March 11, 2005) Nicole DeHuff, of the movie "Meet the Parents", died, at the age of 31, of complications of pneumonia. Believe me the literature is packed with all types of bacteria that have caused similar problems. The virulence (misery and disease producing capability) of the bacteria is increasing because we have overused and continue to over use antibiotics.

Let us not forget that antibiotics were not discovered by a chemist, working in a laboratory, mixing reagents together, in a manner akin to the discovery of vulcanized rubber or plastic. Antibiotics are naturally occurring microbial products! They have evolved through a natural selective process of their own to increase the survivability of their developers. Penicillin, for example, is naturally made by a fungus, or mold called *Penicillium chrysogenum*. Over millions of years of natural selection it just happened to develop a defensive mechanism that would discourage bacteria from picking on it. A scientist named Fleming, in the 1920's, just happened to make the correct observations when given the opportunity, that led to its later use, therefore ushering in the antibiotic age.

All antibiotics are technically, naturally occurring (synthetic compounds like *sulfonamides* or *imidazoles*, that have anti-bacterial properties, really should be referred to as *chemotherapeutic agents*, but they're not). Another one, called *streptomycin*, was the first antibiotic discovered by a random screening of soil organisms. After the original discovery of a core of these agents, chemists entered the picture, synthetically expanding both the number and effectiveness of the compounds.

So why is this over use of these agents so bad for us? How could these little creatures all of a sudden figure out ways to thwart these compounds after centuries and centuries of not being able to? The answer lies in the natural selective process itself and the fact, as stated earlier, that there are so dang many bacteria! The human population on the earth is huge, too, so therefore there are lots and lots of people taking antibiotics, all the time. We're all little walking Petri dishes. Every time antibiotics are used, a *selective pressure* is placed on the population of bacteria in a given organism (person). You have many different types of bacteria living on or in you. They all are exposed to the antibiotic. Usually only one or two types are actually causing problems, due to their overgrowth in a certain area. But you take this stuff that exposes them all to it (one of the reasons why diarrhea is a common complaint during antibiotic therapy is because gut bacteria die in huge numbers and water flows into the colon due to what is called the osmotic effect, and then out it goes!). Very large numbers die and hopefully the bad ones left over can be rid of by your immune system and you will recover.

The crazy thing is that because there are such large numbers and also due to the fact that they are always having babies and passing on progeny, they are very likely to change their genetic makeup (mutate) in such a way that they may not be affected as much, or at all, by the antibiotic being used. So in a given population that is constantly dividing, most are killed off but some may have mutated in such a way to make them immune to the agent used. This mutation has conferred on them survivability and they pass on the trait to the subsequent generations when they divide, which allows this resistant strain to become more prevalent in the population and before you know it the antibiotic becomes ineffective and you have to use something else against that particular bug.

Now throw in the fact that not only do we treat humans with antibiotics but their overuse extends to animals, too. We've even come to feed our more expensive cattle, chickens, and dairy cows antibiotics prophylactically. We don't even wait till they are diseased. We just give it to them to make sure they don't get anything! This allows them to be reared in a filthier environment, too.

Why in heavens name would we do crazy things like this!? The reason is simply because a sick animal costs money to make better and healthy animals grow quicker and can be butchered earlier thereby improving the bottom line as much as possible. The shelf life of meat, eggs, poultry, and dairy products is also lengthened. <u>They do it to maximize profits!!</u> They don't care if they are contributing to the ineffectiveness of antibiotics. They want their money and screw the rest of us. They say they do it to protect us and keep prices down. They slap some lipstick on that pig and sell it to us. Then when we get super bugs like the toxin producing *E. coli 0157:H7* that caused the Jack in the Box scare last decade or antibiotic resistant strains of *Salmonella* in chickens and their eggs, they just throw up their hands in wonder and ask why a thing like that would happen.

As big a problem as this livestock approach is it pales when compared to the problems associated with antibiotic misuse as it pertains to the "Black Market" worldwide. This is especially evident in countries lacking any semblance of a Public Health system. Former Soviet Republics fit into this category and are responsible for the relatively recent development of virulent *Diphtheria* strains and drug-resistant *Tuberculosis* strains.

Oh, if it was only as simple as just this. To throw a monkey wrench in the works further, these little microbes aren't just happy with the DNA they have. They actually covet any DNA that is just floating around and because of this they are able to "acquire" gene sequences that are available and incorporate them into their package of gene sequences. Some of the gene sequences code for a protein that interferes with a particular antibiotic and because of this the bug may acquire the ability to be less sensitive to the antibiotic or even completely insensitive to it (without even being exposed to it beforehand). Bacteria share the "blueprints" for resistance! This occurs between different *species* of bacteria, too, not just of like types. Groups of genetic material, transferred in this way, are called *transposons*.

They have another way of acquiring "blueprints" to thwart antibiotics. This involves more organized packets of DNA called *plasmids*. Most if not all bacteria have their *chromatin* (DNA) organized and sequestered in one part of their cell compartment (although it is not surrounded by a nuclear membrane as with our cells). Sometimes they have an accessory package, the plasmid, they are willing to swap, or give a copy of, away.

I should also include the most common way they have of swapping DNA and that is by *sexual conjugation* (bacterial intercourse). The result is the same (increased resistance likelihood) and together with their tendency to mutate they have adequate abilities to make our lives miserable.

Further muddying the waters, these bacteria, through shear numbers and constant trial and error due to the selection pressure of the overuse of antibiotics, have "invented" different ways of frustrating successful antibiotic use. For example, penicillin is effective because it interferes with construction of the cell wall of bacteria leading to the bug's death. Bacteria that are resistant have an enzyme that cuts a critical part of the penicillin compound, its *beta-lactam* ring (hence the term *beta-lactamases*). Penicillin becomes less effective or ineffective, the cell wall is completed, and bacteria survive. There are lots and lots of cell wall inhibitors in our antibiotic arsenal, some naturally occurring and some semi-synthetic (chemically manipulated natural agents). All have experienced problems with resistance.

Some antibiotics, like tetracycline, interfere with the bacteria's ability to make protein. They do this by disrupting normal ribosome function. As alluded to in the Nitrogen section, protein is made in all cells on ribosomes. On these protein-aceous protein factories, mRNA that has been transcribed off the DNA, is translated into protein by using tRNA (transfer RNA), amino acids, and ATP (energy). Antibiotics such as tetracycline disrupt this protein making process and if the bacterium cannot make protein it cannot conduct normal daily operations and it dies. Bacteria that have become less sensitive or resistant to this type of agent have simply changed the area where the tetracycline binds to the ribosome and therefore less, much less, or no tetracycline binds. This trait, it is thought, occurred by a simple random mutation that conferred such survivability to the host that it was passed on to progeny, others by sexual conjugation, others still by transposon and plasmid swap and common resistance ensued.

Here's an analogy: you have a spotter at a busy intersection that is instructed to not just spot (no selection pressure) but to flatten the tire of any Volkswagon of any type of any year (a reasonably strong selection pressure). He does this by identifying the vehicle by its VW emblem on the exterior of the car and then magically (it's a hypothetical situation) causes a hole to form in any of the four tires. Occasionally, a VW that has been vandalized, say, or has fallen into a state of disrepair, and has no emblems, drives through unscathed. Well, word gets out pretty fast and after a while no one sports an emblem any more and there is no effect. Is this exactly the same? Hardly, it's just a visual to relate to.

A more recent and more ominous turn in the resistance game observed by researchers and practitioners is the development of "energy driven pumps" that specifically focus on antibiotic compounds that enter the bacterial cell compartment. Incredibly, these proteins have the ability to latch onto the antibiotic, escort it to the cell wall, and eject it from the bacterial compartment into the external area. "Like microscopic bouncers protecting clientele from undesirable riffraff", according to Laurie Garrett, author of *The Coming Plague*. And there is no end to their creativity. Every time we invent, they counter the move. Then they tell all their friends and resistance increases.

It all gets back to man's arrogance. Thirty years ago some scientists and researchers were actually predicting the end to all disease. Along with smallpox, more if not all diseases would be eliminated, supposedly, in thirty years. We were so sure of this that pharmaceutical companies left the "sure thing" to others. New researchers focused on other fields and the available money for research went elsewhere. During that time things just got more and more difficult. We're now further away from that prediction than when it was made. This is not only because we have advanced far enough scientifically to see how silly a prediction it was, but also because of our misuse of antibiotics and the monsters we've created.

It was in the wealthy and medium-income countries that misuse and overuse occurred (and continues to occur) most often. It was also in the same countries that resistant strains most commonly emerged. Ironically, though, it was and is the poorer nations that pay the dearest price because they can least afford alternate drugs. The world is a smaller place these days due to air travel, so bugs spread around the globe very quickly and easily.

So the answer would seem to be to just make newer and more clever antibiotics. Unfortunately, it's not that easy. The development of new antibiotics is very costly and if a company should take on these costs and develop a new, effective drug that saves lives but is very expensive, they look like grim reapers if they try to charge big fees to dying patients. It's a public relations nightmare. Couple that with the likelihood that resistance will occur sooner rather than later and one can see why few drug companies are interested.

So what can be done? As stated in the Hydrogen section we've thrown the world out of balance in many ways and the ramifications of this arrogance and lack of respect for the ecosphere is going to be severe. With the population of the earth ever increasing it is just a matter of time till we get a natural correction and bacteria are likely to play a part.

Viruses

Viruses are amazing little entities. Depending on how one defines life, it is very debatable whether these little conglomerates "live". They don't eat food or release a waste product. They have no ability to derive energy on their own and in fact can't reproduce on their own (generally speaking). What they can do is invade or gain access to certain cells and once this is done they can insinuate their demands for protein synthesis into the host cell and essentially take over. They, in essence, evade the defenses of the host cell, slip in through "the back door", and hijack the cell for their own reproductive necessities. They make protein and then using their newly formed protein specific to their needs, duplicate their genetic material and package it in structural proteins (new *virons*), break open the cell and let all their "babies" out. It's like *Alien* on a tiny scale. They are very good at what they do, as everyone knows.

The number of different viruses a human being may encounter as a disease process is both frightening and staggering. *HIV, herpes, cytomegalovirus, hepatitis A, B,* and *C, influenza, rhinoviruses* (colds), *mumps, measles, rubella* and *rabies* all sound somewhat familiar, don't they? But there is a veritable cornucopia of diseases caused by viruses that most people have never heard of. *Reoviruses* that cause gastroenteritis, *Flaviviruses* that cause *Yellow fever, Dengue* (breakbone fever), and West Nile sickness among others, *Arena* and *Filoviruses* that cause *Lassa fever, Machupo hemorrhagic fever, Ebola* and *Marburg disease*, to name just a few. You really wouldn't want to see the complete list of all the viral ailments you could get because it alone may make you sick. Amazingly, though, they are all around us, just like bacteria. They are much smaller than bacteria (by a factor of 100 to 1000) and they are unaffected by antibiotics.

The word virus means "poison" in Latin. They were originally called "contagious living fluid" (*contagium vivum fluidum*) by one of the two people credited with critical early work in the field. The researchers, a Russian named Dmitri Ivanovski and a Dutchman named Martinus Biejerinck, worked separately just before the beginning of the 20th century. They both worked on the same virus called the *Tobacco Mosaic virus* (it was large for a virus and relatively easy to study).

Biejerinck, educated as a Chemical Engineer in Delft, Holland (the same town in which van Leeuwenhoek made his discoveries) is generally not given as much credit as Ivanovski in the discovery of the virus mostly because he did his work 6 years after Dmitri. Martinus did not have the advantage of reading Ivanovski's work, though, since it was published in an obscure Russian journal. Also, he not only did the same isolation work as Ivanovski, using smaller and smaller filter sizes (to rule out the infectious agent being a bacteria) but he carried it a few steps further by reinfecting other plants with sap (hence the term "contagious living fluid") and using denaturing modalities like formalin and heat to further define the nature of the beast.

Still researchers did not know for sure what the heck this little contagious agent was until the advent of the *Electron Microscope* in the 1930's. Then they could actually see the virus. Also, it was in 1935 that a researcher named W. M. Stanley isolated and crystallized Tobacco Mosaic Virus. He showed that if the crystals were injected into the tobacco plant the plant would get the characteristic disease.

People tend to think in their usual anthropocentric way (all life revolves around humans who are the greatest of all creatures ever) that viruses only affect or infect us. Wrong, wrong, wrong! Viruses have evolved over eons to not discriminate. There are viruses that affect and infect all species of plant and animal on earth. We tend to know more about the ones that infect us due to the urgency created when humans suffer and die (not to mention the fact that we're overpopulated so there are lots of targets for the viruses). We don't care nearly as much about a virus affecting a plant or animal unless it is a plant or animal we have "domesticated" and a corporation suffers a loss of capital as a result...then there's action. Hey, even bacteria are picked on by viruses. They are called *bacteriophage*. Recently they have actually been used in research to add genetic information to certain bacteria.

One virus that was not mentioned above is the virus that causes *smallpox* (*Variola*). Many know it well because it was a disease that most people 35 and older were immunized against when young, not to mention that it has been all but eradicated since 1977 (last naturally occurring case in Somalia, 1977). It had truly been a scourge since antiquity. In fact, it's first suspected outbreak, at least 3000 years ago in China and India (according to written record and mummified remains), was only the start of a co-history with the virus punctuated by epidemics that came as often as every 5–8 years. Luckily it was generally lethal in only 30% of cases but American colonists and their offspring generally suffered a much higher death rate, that is until the "miracle" of vaccination.

At first they didn't call it vaccination for reasons soon to be revealed, but instead they used the term *variolation*. This is because the actual virus from an active lesion of a person with smallpox was used. By saying this I mean that either pus from a smallpox lesion was introduced into a cut on a previously unexposed person or scabs from a lesion were insufflated (breathed in through the nose). This usually resulted in a milder form of the disease with a death rate of only 1% and it made the person immune to it for a number of years!

Incredibly, people in China first did this, according to written record, at least 1000 years ago. The technique slowly made its way through the Middle East until Lady Mary Wortly Montagu, the wife of the British Ambassador to the Ottoman Empire, learned of it in 1717, while in Constantinople. She had survived smallpox as a child and apparently had a son variolated there at this time. In 1721, back in England during an epidemic and at the urging of Lady Montagu and the Princess of Wales, some prisoners and abandoned children were experimented upon and variolation was found to work. Later that same year, in the Colonies, it was done for the first time at Harvard. It was not without its risks though, for many died, including the son of King George III (I would've hated to be the "Doctor to the King" during that shift).

It wasn't until, in the 1790's in England, that <u>vaccination</u> started. Edward Jenner observed that for some reason, milk maidens (women that made a living milking cows, touching udders and sharing viruses with cows) were unaffected whenever there was an epidemic of smallpox. He reasoned correctly that a similar disease in cows somehow conferred immunity on the maidens (in essence, vaccination). He went so far as to isolate the pus from active cowpox lesions and, once again using orphans, proved that it made people immune for several years. This was the first time that vaccination on a large scale, with very, very few deaths resulting, was done.

So, oddly, the term, vaccination was used only because the source was the cow (from latin, cow-*vacca*). The term continues to be used when promoting immunity to just about anything, these days, and cows aren't even remotely involved (vaccines are now used that utilize only dead or weakened bacteria or viruses).

The last question to be answered here is why the heck was some respected physician going around willy-nilly conducting experiments, potentially lethal experiments, on children?! I guess it was just a different time back then, huh? With life expectancy on the average only in the forties the value of a life was different so certain people became expendable (where was the ACLU then!).

Humankind's desire to protect itself from it's hostile surrounding enabled man to technologically improve through the Stone Age, the Bronze Age, the Iron

Age, through the Industrial Age to the Computer Age. It is true that many advances occurred in order to protect humans from each other, but as sad as that is we have advanced nonetheless. We have better, more efficient, more refined methods of doing everything! Medicine and science has kept pace with our advance. In many ways our industrial accomplishments have ushered along improvements in science and medicine and with that bacteriology and virology. Particularly poignant examples of this are the advances made during the Spanish-American War and the creation of the Panama Canal, the latter a marvelous feat of engineering. Due to the type of work and more importantly the location (both Cuba and Panama feature warm swamp infested areas), humankind's collision with the virus that causes Yellow Fever occurred.

Yellow Fever had been a formidable disease long before the Americas were settled by Europeans. It had a reputation of bringing death and misery regularly in locales that were warm and that had stagnant water areas (usually in the form of swamps) nearby. With the settling of the New World and the inhabitation especially of the Gulf coast, Yellow Fever became a scourge. When outbreaks occurred in port cities, the "Yellow Jack" went up (a yellow flag recognized by all to indicate an epidemic and to steer clear). All commerce would stop and people would be quarantined who showed signs of the ailment. Between 1817 and 1900 Yellow Fever outbreaks occurred nearly every year along cities and towns of the Southeastern U.S. and the Gulf. The epidemic in New Orleans in 1853 killed over 9,000 people alone.

France had wanted to build a passage through Central America, as a short cut to the Pacific, for years and in 1878 concessions were made by the Colombian government (Panama was a part of Colombia then) that enabled France to try. They started their attempt in 1882 but were frustrated by the diseases encountered as well as technical problems. After several more attempts failed they gave up in 1894 and relinquished rights to the United States. The U.S. had wanted to do this for a while and had actually sent a surveying crew to Nicaragua, in 1887, to see if it was feasible there. When they acquired the building rights from Nicaragua that site was chosen over Panama and work was started in 1889 but a stock panic drove investors away in 1893 and work was halted there. In 1897 a fact finding Canal Commission was appointed by Congress to restart the process and again the Panama and Nicaragua sites were considered. Again, Nicaragua won out and the plans were set to be signed by President McKinley in 1901, but he was assassinated. His replacement, Teddy Roosevelt, started an era of strained relations with Colombia which led to the independence of Panama as they broke away from Colombia with the protection of the United States. This opened the way for the Panama site and in 1904 the problem of disease, especially Yellow Fever and Malaria, was addressed. The U.S. was not going to repeat the misery experienced by the French between 1882 and 1885 when they lost an estimated 20,000 workers to Yellow Fever.

It was about the same time that the Spanish-American War occurred (1898 "Remember the *Maine*") and although there were less than 1,000 battlefield casualties (not counting the *Maine's*) there were triple that number lost to disease, especially Yellow Fever. The Army Medical Corps stationed a medical team headed by Dr. Walter Reed, in Cuba, to find out what caused Yellow Fever and to find a way to control it. Congress did not want troops stationed there if they were just going to get sick.

On June 21, 1900, work by the team started in Cuba, during an outbreak. There were two theories for the cause: the mosquito and *fomites*. Fomites were explained to be something that is transmissible even on bedding and clothing.

The team, by using humans (usually soldiers) as guinea pigs, figured out that it was something transmitted by certain kinds of mosquitoes and they even went so far as filtering it and deciding that it wasn't bacteria. They instituted protocol for controlling the mosquito population (either eliminating or oiling all standing/stagnant water) and it worked. The number of cases dropped to very few. The Panama team implemented this tactic of controlling the mosquito population also and they saw both Yellow Fever and Malaria cases drop to very low levels. Work then proceeded with just technical problems to frustrate the engineers but by August of 1914, the Panama Canal was completed. A modern marvel made possible by the work by Dr. Reed in Cuba and implementation of the protocol in Panama. Yellow Fever had been beaten.

So what the heck is a virus? As troublesome as they may be for all forms of life they are nothing more than a nucleic acid molecule surrounded by amino acid chains with a very finite number of extraneous support molecules; in other words DNA or RNA (genetic material) with a little shell (structural proteins) to protect it and some proteins (nonstructural) to initiate different transcriptions off the genetic material (which ultimately results in protein manufacture in the host cell). They've merely evolved through the eons to be able to respond to their environment in such a way that enables them to make large numbers of copies of themselves. They come in some freaky, science fiction type shapes, relatively simple geometric shapes, polyhedric shapes, and even some shaped like a section of galvanized pipe (tobacco mosaic virus). Whatever their shape their goal is the same; find the preferred host cell type, take over, and make more copies of themselves.

As one would figure, understanding that diversity is biology's way, there are many different strategies of finding the preferred cell type. Yellow Fever is an *Arbovirus* (from A̲rthropod bo̲rne), which means it has evolved a strategy of using an Arthropod (insect, in this case mosquito) vector. Many viruses have adapted to use many different insects in their quest for their favorite cell type.

The Flavivirus that causes Yellow Fever can live and multiply in certain types of mosquito; in their gut and in their salivary fluid. Prior to reproducing the female must have a blood meal or two and when a human is bitten (or poked), some of the mosquito's saliva ends up in the human and there are many, many viruses exchanged (because they are very small and very numerous). The effects may be felt in the intestinal tract to some extent but mostly the liver is affected and death can result in 3–6 days. The Yellow Fever virus prefers human liver cells and a vaccine is available. West Nile Virus does things very similarly to Yellow Fever as does Dengue Fever (as well as *St. Louis encephalitis* and *Japanese encephalitis*) in their transport via mosquito saliva. The preferred cell type though is different (bone and joints for Dengue and the brain for any encephalitis).

Taxonomic classification (organization according to similarities and differences) of viruses is very complex. It is compounded by the yearly discovery of more and more of them. Things change all the time due to new discoveries and similarities and differences. When I went to school and took Microbiology in the late 1970's, the Hepatitis viruses were classified together. They were originally lumped together because of the preferred cell type infected (liver). Now with the incredible biochemical work done and the advances in ultrastructural anatomy Hepatitis A, Hepatitis B, and Hepatitis C (what used to be non A/non B Hepatitis) are in completely different families.

Hep A is in the *Picornaviridae* family. It is very small and its nucleic acid inside the protein coat is single stranded RNA. Its cousins in the same family are the virus that causes *Polio* and the common cold!

Hep B is in the *Hepadnaviridae* family, is larger and more complex than Hep A and has double stranded DNA. One of its cousins is what has come to be known as *Hepatitis D* or *delta virus infection*. Delta virus infection cannot occur without either Hep B to help it or another cousin from its family (*woodchuck, ground squirrel,* or *Pekin duck virus*).

Hepatitis C is a member of the family *Togaviridae*. It is even bigger than the other two Hep viruses (almost twice their size) and has single stranded RNA inside its coat. Its only cousin is the virus that causes *Rubella*. This virus was not even discovered until 1989. Yet it has been around for decades frustrating practitioners. Technologically our ability to detect it had to reach a certain level of sophistication. The same scenario occurred with AIDS and the detection of the causative virus (HIV or human immunodeficiency virus).

AIDS appeared to emerge out of Africa based on the fact that the Green Monkey population in sub-Sahara Africa had antibodies to HIV. This could only have occurred if the virus, in the distant past, wiped out most of the Green monkey population except a small number that were not killed by it. Instead it just caused a much less serious infection in the few survivors and the population reestablished itself from this subset of less affected individuals (with antibodies).

At some point, since Green monkeys are hunted and eaten by different tribesmen, someone probably ate incompletely cooked meat or was bitten in a scuffle with one of the creatures infected with HIV and the whole mess started from there.

It happens to be a virus that can only be transported by the blood or semen route so has become a STD (sexually transmitted disease) that is exhausting National budgets in Africa and elsewhere.

In the United States it first was noticed among the homosexual population in the late 1970's. At the time the CDC (Center for Disease Control) had its hands full with an outbreak of *Legionnaires' Disease* (caused by a previously unknown bacterium called *Legionella pneumophilia*) and the prospect of an influenza scourge that ended up never materializing. When the Reagan administration took the reigns of government in the early eighties, budgetary outlays were channeled elsewhere for the most part due to the conservatives being unconcerned with a disease that was killing gays. Finally, due mostly to the contamination of the nation's blood supply as evidenced by a steep increase in the new cases of AIDS among hemophiliacs, more grant money was allocated for research. This led to the discovery of the causative virus and a strategy to combat the disease that continues to evolve and improve.

HIV is in the family *Retroviridae*, has RNA in its core, and has a cousin that causes *lymphoma*. It works by attacking T cells (an important part of the body's immune response) and the result is an immune system that doesn't work properly. It is evidenced early on by the appearance of normally easily controlled infections (*Candidiasis* or yeast overgrowth, *Pneumocystis carinii* pneumonia, and *cytomegalovirus* to name a few) and an odd lesion (*Kaposi's Sarcoma*) that is almost never seen otherwise. If left untreated these and other normally benign infections can kill.

Treatment involves drugs (antiviral agent) that interfere with the specific way that HIV makes protein. They can be very effective as illustrated by the seemingly normal lives led by some (Magic Johnson for example). There are currently two main disease types: a virulent strain active in urban centers throughout the world and a less lethal form that is seen throughout sub-Sahara Africa.

Another virus that has been around much longer than its "discovery" date indicates is *Hantavirus* (also referred to as *Korean Hantaan*). First isolated and confirmed as a virus (around 1970) by Dr. Ho Wang Lee of Korea University Medical School in Seoul, it caused quite a problem during the Korean conflict (between 1951 and 1954). 121 GIs died of the mysterious ailment and more than 2,500 were taken ill (an unknown number of Korean soldiers were affected).

As detection abilities became more sophisticated, 11 other forms of Hantavirus were discovered in Eastern Europe and Asia. The human kidney was the preferred cell type of the virus and the reservoir for the virus was found to be the wild rodent population.

Oddly, in 1993–4 an outbreak occurred in New Mexico among the Navajo Indians there. This was found after extensive study to be another form of Hanta, a form that in addition to the kidneys also affected the lungs. This discovery was associated with what became known as *ARDS* (acute respiratory distress syndrome). The other thing that changed was the lethality. Instead of a 10% or so rate of death the new form of Hantavirus had a 70% rate! The reservoir for the new type was confirmed to be rodents of the family *Peromyscus* (deer mice) and especially *P. maniculatus.*

Detection abilities continued to improve and the virus's range was found to include lots and lots of different rodents including voles. The really deadly form though was and continues to be confined to the deer mouse population. Other strains of this virus found throughout the world and in the U.S. (in the other rodent populations) produced either a much more mild form with a lower death rate or no detectable disease at all.

It now has been postulated that the mysterious disease called "trench nephritis" that sickened and killed many soldiers during World War I and the American Civil War, was Hantavirus. Like a number of other diseases (Machupo hemorrhagic fever in S. America for one) it lives in the rodent population and when the feces or urine is deposited in areas that humans inhabit, the virus that is in the waste may become airborne and inhaled, thereby starting the disease.

The improvement in science generally and biochemistry in particular, has been impressive in the last 100 years but the improvements of microbiological pathogens' abilities to cause suffering and disease has allowed viruses to become more and more of a scourge as our human population gets larger and larger. This is because viruses live in numbers that we cannot even conceptualize. Humans have trouble with anything over 100,000. The number becomes a blur after that.

The national deficit in the trillions is way over our heads. A trillion is only a one followed by twelve zeroes. Biochemists deal in huge numbers since atoms, and their bioactive conglomerates, are so tiny. A dozen has come to be known as twelve to the average person, but a *mole* is a 6 followed by 23 (visualize that) zeroes to the average chemist! Microbiologists and specifically virologists live more on the chemist's side of quantities when studying their prey.

To drive home the point lets take the virus voted most likely to decimate the human race and try to conceptualize the numbers problem. *Influenza* and its pandemics (worldwide epidemics) is the subject. The number of virons in any given area or region is astronomical, true, but <u>what we're talking about here is not</u> <u>something that is regionally restricted due to the type of animal or insect reservoir</u> <u>it exists in</u> (mosquitoes' life cycle, activity, and warm weather requirements, limits Yellow Fever, Malaria, Dengue, and the other mosquito-born diseases to temperate areas of the globe and only during rainy seasons). <u>Nor is it restricted by</u> <u>transmission requirements</u> (blood exchange and sexual transmission requirements of AIDS, Herpes, and many others). <u>Influenza infects in all climates and regions</u> <u>and only requires humans to interact or frequent the same places</u> (as everyone knows inane things like door handles can spread the flu) <u>to infect</u>.

Its communicability is crazy! So it not only enjoys the powers of these dizzying numbers to insure its regional transmissibility, but also has an across the board monopoly on victims (especially because of air travel and the "smallness" of the world now). Compound that with the regular tendency to change just enough to evade our immune systems and you have one precooked recipe for disaster. For, you see, it is actually unfortunate that influenza virus as with any virus is not perfect. Natural selective processes govern its change. Through antiquity and before, they have, through trial and error, honed their abilities. A well adapted virus does not kill all of its hosts. It would have no where to go. So it adapts by killing a variable percentage of, say 15% and it secures secondary reservoirs in, say, pig and fowl populations (in fact it is thought that these influenza strains first appeared in pig or fowl populations due to close proximity from domestication of these species).

But what if the virulence (lethality), by a random biochemical quirk (random mutation), was unusually high and the result was a 50% death rate. Couple this with immune systems gone flabby from the recent (generational) lack of an immunity-conferring pandemic and you will have an absolute worldwide catastrophe. The likelihood of this happening, given how smart we are, is difficult to estimate but as <u>all</u> populations, animals domesticated to feed us <u>and</u> humans themselves, increase, the world will probably reach a point where something would have to give.

Influenza is a member of the family *Orthomyxoviridae*. Because of how much it concerns us, we have spent a lot of money studying it and we know it pretty well. It's big for a virus, nearly three times bigger than Hepatitis B (still it's only one tenth of a micrometer). In its core is single stranded RNA. It's circular or oval and has a very well studied protein coat.

In immunology it's almost always about the protein coat or outer capsule (and what is poking out of its shell). The way the viron gets into or penetrates the host cell is determined by enabler proteins imbedded in the capsule. The most prevalent and important are two enzymes called *neuraminidase* and *haemagglutinin*. They, by their positioning, numbers, and biochemical tendencies, allow the evasion of host defenses, fusion with the cell, and entry inside. They are both just big proteins with some sugar mixed in (hence they are classified as glycoproteins) and as a Volkswagon Beetle has a characteristic shape, so do these molecules. That's where our immune system comes in. We have the capacity immunologically to make antibodies that are specific for the locations of these glycoproteins located on the shell. In order to do this we must be exposed to the virus and then in a thoroughly studied way (that is still not completely understood) the virus is read as a threat and antibodies are formed. If you are strong enough or supported medically enough by the time your immune system overwhelms the invaders, you live and hopefully completely recover. If not, then you expire and the virus moves on (it's not very caring).

So the changes most talked about by researchers are the position of the two enzyme molecules, neuraminidase and haemagglutinin. These are designated as antigens, or targets for the immune system. Over time the change in location on the surface is referred to as *antigenic drift*. Antibodies formed in the past don't grab on like they used to and the entire immunizing process must happen again to be safe (confer immunity). These antigens are so important that the locations are referred to in the nomenclature of each strain. So *A/swine/Iowa/3/70(N1H1)* is shorthand for a type A virus (there are four types A, B, C, and thogotovirus), whose original host was the pig, was first discovered in the geographic area of Iowa, is strain number 3, and the year of emergence being 1970, and then the designation in parentheses of the location of the antigens.

The H and N antigens seem to change every 10 years or so with severe changes that could cause a pandemic every 30 years or so.

So to recap the crux of the matter, viruses are everywhere. They have evolved a number of strategies of persisting and infecting. They affect and cause trouble in a wide variety of plants and animals. There are unfathomable numbers that to humans are inconceivable. We are smart but not smart enough. As we arrogantly limit the earth's biodiversity by increasing our own human population and the very limited types of animals and plants we've domesticated, and destroying niches of animals and plants we have little use for, we create the scenario for the plague of plagues. It's not a matter of if but a matter of when. The world is self correcting and if we don't take steps to reestablish certain equilibriums and promote biodiversity there will be another round of mass extinctions, this time possibly involving the species that caused it.

The Earth and Geologic Time

If you were to look out at the earth from the perspective of wherever you may be and describe what you see, invariably you would describe the vegetation that grows upon the earth in your area or possibly animals that frequent the area you reside in or maybe you would describe water in some form and its relationship to the shore and to your surroundings. You would be inclined to describe the earth itself only if there was something different such as a snow-capped mountain, an odd gouge out of the earth, or maybe a confluence of several terrain types.

Ask a geologist to describe the earth and you could never get him or her to shut up. Moraines, gabbro, pillow lavas, gneiss, granite, quartzite, and jasper conglomerates are just a few terms that would roll off the tongue of a geologist primed to describe his domain. But I challenge anyone to engage in a discussion with a geologist and avoid even broaching the subject of time or more accurately "geologic time". It can't be done.

Just as the concept of large numbers of atoms or molecules (as discussed in the Virus section) is over the head conceptually of humans so to is the incredibly large numbers of years considered in Geology generally and geologic time specifically. The Sierra Nevada Mountain Range, being a "recent" event geologically, is still the result of tectonic movement in the Jurassic era, *200 or so million years ago!* The concept of "200 million years ago" is difficult to fathom because as organisms with our "recently" acquired ability to reason, we didn't need it. Humans, during the last 2 million years were hunter-gatherers that could conceptualize (our advancement that sets us apart from all other creatures, the ability to reason, was thought to have evolved in us during this 2 million years) timeframes of only a few generations back and a few forward. 50 year concept windows make it difficult to relate to geologic time.

To put this concept in proper perspective, if the age of the earth was considered in terms of a calendar year then basic one-celled organisms didn't even appear till Halloween (~650 million years ago). Dinosaurs appear on the thirteenth of December and survive only to see Christmas. The last glacial period ends at 11:59 PM (11,000 years ago) and the rise and fall of the Roman Empire takes all of 5 seconds!

It's amazing that we can be the newcomers to the earth but even though we've come to dominate in only the last second we think its all there for us to do with what we want, as if it were preordained.

It's all well and good to make suppositions about time and throw "millions of year" era names like *Jurassic* and *Silurian* around but how exactly do we know how much time has gone by since a particular rock has formed, or a particular animal, whose "fingerprint" is encased as a fossil, lived? The answer lies in what is referred to as *Radiometric Dating.* As discussed briefly at the end of the Carbon cycle section, *carbon-14 dating* is considered a radiometric dating tool, but of once living things whose lives were centered around carbon (all species of life can be dated this way, given their age some more accurately than others). Due to the advent of computers and assay techniques (utilizing diffractometers, spectrometers, and electron probes) that can measure "small" amounts of atoms and their isotopes (and their proportion relative to each other) carbon-14 dating and other radiometric tools can be extremely accurate.

The accuracy of these tools lies in the consistent and predictable nature and behavior of atoms, their isotopes, and molecules that contain them. Some isotopes of just some of the elements of the *Periodic table* are *radioactive.* This means they are not stable enough to just exist, but are compelled by a lack of "balance" at the nuclear level, to give off energy in some form, as they disintegrate to a stable situation.

Picture an overloaded high-speed washing machine that is unbalanced. I have seen what those babies can do, the racket they create, in just a minute or so. What if an unbalanced washer was plugged in, kept unbalanced, and allowed to go on, and on, and on. Something would ultimately give. Something would reach its elastic limit and break or the thing would just release energy as it overheated and caught fire.

Well, with unstable isotopes, the energy given off isn't manifested in such a macroscopic occurrence like fire. Rather, radioactive isotopes undergoing *decay* give off or emit "particles". Generally what is given off is either an *alpha* particle (which is a *helium atom*), a *beta* particle (an *electron*), or a *gamma* particle (a high energy *photon*). The emission rate of whatever the isotope and whatever is emitted is constant through time. This radioactive decay "schedule" is incredibly precise no matter what the sample. Carbon-14 has what is called a *half-life* that says: of the starting amount of carbon-14, after 5,730 years only half will be left. The other half of the starting C-14 will have changed to the more stable C-12 by way of energy release. Carbon-14 dating is just one example, and is the tool most people have heard of.

In geology the latest and greatest tools use similar concepts but different elements. The tool most often used, due to its consistency and preciseness, when dealing with very old samples, is *Uranium-Lead dating*. It was first used in 1907 but via improved assay techniques and the use of computers it has become very precise. The best part of a rock, at the molecular level, to use with this tool, is what is called *zirconium orthosilicate*. This crystalline structure is found in many old rock types. The reason why it's the preferred part of the rock is not because of zirconium's isotope (zirconium is the 40th element of the Periodic table) that has a half-life of 63 days. You can't do much with a 63 day half-life isotope. The reason is simply that uranium tends to be found in all rock in a certain amount and due to zirconium orthosilicate's ease of handling, and when cut, its low frequency of contamination, it makes the best "holder" of the uranium to see how much of the isotope has changed to lead and how much of the isotope has remained unstable, still emitting energy (which is measured). With this sample type only ten one-millionths of a gram is necessary for an accurate assessment of the age (remember that even though the sample is small, the number of atoms is incredibly large and therefore ample for accuracy). Uranium-238 has a half-life of 4.5 billion years!

Samarium/Neodymium radiometric dating as well as *Argon-Argon* dating and *Rubidium-Strontium* dating are other tools that are used that employ similar principles. The Argon dating tool measures the relative amounts of a radioactive isotope of potassium (K-40) to the amount of Argon-40 (Ar-40) in samples and deals with a half-life of 1.26 billion years. The Samarium-Neodymium system relies on the 106 billion year half-life of Samarium. Samarium (element 62 of the periodic table) is a rare earth element which occurs as a metal in the *mantle* (the intermediate shell zone of the earth below the crust but above the core). When volcanoes erupt and magma comes up from the mantle, it carries with it a certain level of radioactive Samarium which then begins its decay to Neodymium. By use of this tool the date which a magma eruption occurred can be assessed. The Rubidium (Rb-87) decaying to Strontium system has a half-life of 48.8 billion years.

Presently there are over <u>forty</u> radiometric dating techniques employed in this field. Samples can be and in fact always are checked and cross-checked with multiple systems to ensure accuracy. Over one thousand papers concerning radiometric dating are published *per year*. Scientists test and retest everything before publishing. The science that is relied on in radiometric dating is the same science that is used in many other fields and allows the medical and industrial advances that make our lives easier and safer. It is irrefutable and if you don't believe it then you just don't want to nor do you want to take the time to study it.

Once radiometric dating has established a rock's age, then the age of fossils found in the strata (layers of buried earth) where the rock was found, can be inferred. This is only a guess, though, and when fossils are radio-dated, their age can be accurately assessed. The fossil record has shown us that since multi-cellular life began in the *Cambrian Period* some 540 million years ago, species have appeared, existed for a number of millions of years (or less) and then have died out, never to be seen again. This has given rise to the concept and use of *index fossils*. Index fossils were used and continue to be used, when drilling into the earth, to find areas where coal, oil, and natural gas can be expected in the strata, and with great accuracy, due to the age of the strata they are found in.

The concept of index fossils is really quite simple. If I told you that I had a photograph in which Teddy Roosevelt and John F. Kennedy were both present, could you guess, within 2 years, when the photo was taken? Of course you could if you knew that JFK was born in 1917 and Teddy died in 1919. You would know this without radiodating the photograph but could corroborate it with that technique, also.

Without these super-sleuth tools and others like *Seismographs*, *Magnetometers*, *Mass spectrometers*, and *gravitational field strengths indicators* the study of the earth would be difficult. Before the advent of these tools the Hindu fable about the blind men all feeling different areas of an elephant and coming to different conclusions about what was being explored was poignant.

> It was six men of Indostan
> To learning much inclined.
> Who went to see the elephant
> Though all of them were blind,
> That each by observation
> Might satisfy his mind.

Because of the six different areas touched, the six came up with six different concepts of what it was. The one that touched the side thinks it is "a living wall", the one that touched the tusk thinks it "a spear", the one examining the trunk thinks "a snake", the one concerned with the tail envisions "a rope", the fifth touches the ear and conceptualizes "a fan", and the last explores the knee and relates it to "a tree".

And so these men of Indostan
Disputed loud and long,
Each in his own opinion
Exceeding stiff and strong,
Though each was partly in the right,
And all were in the wrong.

Geologists studying different areas a hundred years ago had different ideas about our earth and with good reason. The earth and the rocks and terrain types can be very perplexing. That is why it is so amazing that James Hutton, a medical doctor, wrote the *Theory of the Earth,* presenting it to the *Royal Society* in 1785! It was incredibly insightful and accurate though not without its mistakes.

Theories of how the earth formed, at the time (1700's) that were in vogue, were of course the biblical version and also the concept called *Neptunism.*

An Archbishop named James Ussher, of Ireland, in attempting to date the earth, counted generations back in the bible and arrived at the creation date of 4004 B.C. Other aspects of creation centered around Genesis and geologic changes around the Great Flood.

Neptunism declared that rock of every kind (*igneous, sedimentary,* and *metamorphic* being the three types) had precipitated out of solution from a world engulfing ocean. Depending on when and where it precipitated is what determined its exact type (whether it became gneiss, schist, granite, slate, or sandstone etc). Once it formed it pretty much stayed there unchanged and there wasn't an explanation as to where all the water went.

The lead proponent of this scientific hypothesis was a guy named Abraham Werner. He had explanations for everything and in fact was able to convince most people who did not ascribe to the biblical picture that this hypothesis was fact. He even insisted that volcanoes erupted due to spontaneous combustion of underground coal, and it was believed. So although he had never even left his home area of Saxony in Europe, and only by way of his charisma, oratorical ability, and his burgeoning reputation, he was able to paralyze deductive reasoning and the progress in Geology for some 150 years. He was later facetiously called "a kind of scientific pope" that no one would challenge, by the Director General of the Geological Survey of Great Britain and Ireland, Sir Archibald Geikie (in 1905).

Neptunism and Werner's earth was seen to be compatible with Genesis and the biblical creation of the earth and was accepted by the Pope. This contributed greatly to its acceptance. Hutton's theory, though with its supporters, was deemed a "visionary fabric" and rejected generally. Hutton didn't care because he traveled widely and studied extensively varied terrain types. He saw rock formations and worked backward in time to visualize what must have happened. He had a great friend in a renowned chemist named Joseph Black who supported him and when possible chemically corroborated his findings. The two of them were the founders of the *Oyster Club*, which was a conglomerate of great minds of the time and region. They would meet with the likes of Adam Smith, James Watt, Benjamin Franklin, David Hume, John Clerk, and John Playfair and discuss the science of the day. Hutton's work was often debated and was generally accepted as being brilliant. "Hutton shared with them the developing fragments of his picture of the earth, which, in years to come, would gradually move the human world from a specious position in time in much the way that Copernicus had removed us from a specious position in the universe" (from John McPhee, *The Annals of the Former World*).

The idea that sailors on Columbus's ships believed they would fall off the edge because the world was flat is widely disputed now. It was generally accepted that the world was round in 1492. In fact, since *Pliny the Elder* in the 1st century A.D. it had been inferred. *Ptolemy*, the great Greek geographer, astronomer, and astrologer derived his maps from a curved globe around 100 A.D. (He also advanced the earth-centered planetary system that Copernicus later refuted). Certainly by the 11th century when the Arabic *Astrolab* became in widespread use to navigate by stars, the world was generally accepted to be spherical.

It wasn't until the 1960's that the real revolution in geology occurred. This was the result of many things coming together but the biggest impetus to it all was remarkably the data that came about as the result of the Nuclear Test Ban treaties signed by the major powers. In anticipation of the treaty, hundreds of *seismographs* were placed all over the world, so as to monitor nuclear weapon detonation. The by-product of this was that *seismological events* (earthquakes) were seen to occur much, much more frequently than previously thought.

As early as 1956, oceanographers at Columbia University had suggested that most earthquakes occurred out in the ocean. The data collected by the seismographs confirmed that. It appeared that seafloors were spreading causing deep cracks and mid-ocean rifts. Also, from this seismological data and observations of non-oceanic faults around the world, like the Hayward and San Andreas faults in California, it became roundly accepted that land masses occurred on some twenty-odd crustal segments called *plates* that drifted around the world. The plates were thin and rigid like eggshell and were generally 60 miles or so deep.

The surface measurements were as large as 9000 miles by 8000 miles like the Pacific plate or they were smaller. The plates did not hesitate to crash into one another (as India is doing to Asia) or sideswipe one another (like the strike-slip faults of California) or even move apart from one another (thereby allowing the Atlantic ocean to form). This theory was called the *Theory of Plate Tectonics*.

As soon as cartographers made accurate maps of the world, scientists had been wondering why areas separated by large distances seemed to fit together like a jigsaw puzzle. In 1838, a Scottish philosopher named Thomas Dick published his *Celestial Scenery* in which he commented on this. In 1912, at the Meeting of the German Geological Association, a man named Wegener also commented on this jigsaw-like fit of continents and how they may have drifted from far away to arrive at their present location. He was roundly ridiculed (for 50 years until it was proven, anyway).

While seismographical evidence was a strong impetus toward figuring it all out, *paleomagnetism* was the enigma that fed the fire that ultimately caused its resolution. Paleomagnetism is the study of the intensity and orientation of the earth's magnetic field as preserved in the magnetic orientation of certain iron-containing minerals found in rocks throughout geologic time. The earth has a magnetic field much like a magnet has. As one knows by having used a compass, magnetic north can be figured easily. The central core of the earth though, is made of molten material comprised to a great degree of liquid iron. This liquidity apparently lends instability to the field and every once in a great while the polarity of the entire earth reverses itself. There is general agreement among paleomagnetic geologists that there have been 171 reversals, at regular intervals, in the last 76 million years. One of the quirks of iron containing crystals in molten lava is when cooled to rock, the crystals line up according to the magnetic field being generated by the earth at the time of cooling. They act like little compasses, always lining up with the field, showing forever the direction of earth's poles at that moment in time. In essence, because volcanoes are constantly releasing magma throughout the world, especially on ocean floors, the resulting rock's age can be easily deduced from radiometric dating and their iron-crystal polarity can tell us what direction the poles were in, and where on earth they were "born". As written by Allan Cox in his book called *Plate Tectonics: How it Works*, "the structure of the seafloor is as simple as a set of tree rings and like a modern bank check it carries an easily decipherable magnetic signature".

Using paleomagnetism, scientists have been able to discern that at the end of *Cambrian* time (490 million years ago) the equator crossed what is now a large part of the North American continent in a north-south direction. The area of

Pennsylvania became jungle and swamps clear through the *Pennsylvanian* period (290 million years ago). This is why there is so much coal in the area. The amount is greater than the amount in any other country in the rest of the world.

Back then, when all the carbon-based vegetation died, and layered year after year for millions of years, it was buried, subducted, pressurized and warmed over time.

First it became *peat*, which looks like chewing tobacco and burns similarly. Just as snow is compressed in glaciers resulting in ice, peat is compressed over time into coal (provided there is geothermal heat to cause much of its hydrogen, oxygen, and nitrogen to be given up). Just as ice is about 10 times as dense as snow, so, too, is coal about ten times as dense as peat. There is lots of peat in Pennsylvania (and the surrounding states, also).

Peat that never gets buried far enough, never changes into coal.

Peat that ends up being buried up to three quarters of a mile becomes *bituminous* coal (lower grade coal), over time. In this type of coal, with a microscope, it is easy to see roots, seed coats, bark, and leaves. It is even possible to identify the ancient plants they came from.

If buried deeper and folded severely under pressure, peat, over time, becomes *anthracite* (high-grade coal). It is nearly ninety-five percent carbon and burns very cleanly.

Oil creation by the earth is similar to coal but the requirements are more stringent. Petroleum represents a relatively small percentage of life that lived long ago. In rock, the ratio of all organic carbon to petroleum carbon is eleven thousand to one! The most important factor determining if decaying plant life changes into oil or not, is the temperature. There has to be pressure, also, but temperature makes the key difference. If not hot enough it becomes coal or natural gas and if too hot it burns off leaving combustion gases or spaces in the underground rock.

Natural gas is the easiest fuel to form. Any organic material whatsoever can form natural gas. It just has to die and start decomposing.

The study of paleomagnetism began in the 1920's and through perfection of instrumentation and the incredibly abundant amount of data collected all over the world, plate tectonics and "continental drift" has been corroborated as what is occurring macroscopically over geologic time.

We know now that if you went back in time and looked at the earth at 100 million year intervals, you would see assemblages of continents that only resembled what is here today. The further back you went, the more it would lose that similarity. And if you go into the future you would see the same thing. For example, China and Taiwan are constantly feuding with China wanting Taiwan to become a part of the mainland. If they would just be patient, in another 70 million years or so, it will actually start to merge physically with China. Japan, will be a part of the west coast of North America, somewhere in another 800 million years!

No matter what, for now, the plates will continue to move about on the surface, perpetuated directly or indirectly from the heat at the center of the earth. When the center eventually cools, and it will eventually, then the movement will cease. Then, and only then, will it finally be safe to live in California along major fault lines. But until then people should not be surprised of the occasional earthquake anywhere in the world. They are merely a symptom of this dynamic surface we live on.

Glacial periods or "ice ages" as they have come to be called have punctuated earth's history at regular intervals. Evidence shows that there have been 20 glacial periods over the last 2 million years. It is thought by most experts that we are presently still in an ice age but are actually in a "brief" (brief in geologic time scales) warm period during which glaciers have been receding.

The reasons why glacial periods occur point to the eccentricity of the earth's orbit around the sun as well as variations in the tilt of the earth itself. Also equinoxes (when the sun is positioned over the equator occurring March 21 and September 21 now) vary regularly due to a "wobble" in the earth's rotation that occurs at regular intervals. This is much like a top that while spinning, starts to slow down and begins to wobble. This, as with tilt and eccentricity of orbit, affects the quality and quantity of radiation reaching the earth from the sun. A periodicity of about 22,000 years in glacial periods is the result.

An additional factor that is very important is the location of land masses through plate tectonics and continental drift. When land masses are at or near either one or both poles and other factors as just discussed are ripe, glacial periods occur readily.

Glacial periods are especially important because of the effects they have had on the topography of our planet and also due to the fact that they have had a stimulating effect on humans' colonization of the globe. This latter point cannot be underestimated. If not for the last great glacial period 12–18,000 years or so ago, the North American continent would've had to wait a considerable amount of time before being truly discovered. Also, some 22,000 to 28,000 years previous to that (about the time of the next earlier ice age), Indonesia, island groups around Indonesia, islands south toward Australia and Australia itself, according to archeological data, were settled.

The reason for this is simple. Glacial periods on the planet bind up a great deal of the earth's water in large ice masses. This leads to a considerable decrease in the levels of the great oceans (hundreds of feet). People could actually walk from the Eurasian mainland to many of these areas. Also, great distances of hundreds of miles that now separate land masses between large islands and smaller "stepping stone" islands were decreased considerably thus providing an impetus to seafaring.

In fact, the earliest evidence of watercraft use in human history, predating by 30,000 years other areas (Mediterranean), was found through early human occupation of Australia and New Guinea. It didn't stop there though because abundant evidence has been found indicating islands off Australia (New Britain, New Ireland, the Bismarck Archipelago, and Buka in the Solomon Archipelago) were settled around 35,000 years ago.

So probably because of the initial expansion of humans into this area of islands that were connected then by land, and the ability of humans to see not so distant land across brief expanses of water, did seafaring become an inescapable way of life. As seafaring improved, bolder expeditions across larger expanses of water occurred, whereby, over thousands and thousands of years, all of the South Pacific was settled. The last of the remote islands of Pitcairn island, Henderson island, and Easter island having occurred relatively recently around 800–900 A.D.

The abilities of humans to migrate via the Bering "land bridge" (connecting for a brief period easternmost Eurasia with westernmost Alaska), that would've been covered with ice or would've been dry land, thereby allowing passage of brave groups to the North American continent, was made possible by this last glacial period. The oldest unquestioned human remains in the Americas were found in Alaska and dated around 12,000 B.C. No settlement or archeological find in North America predates this. That is to say that of all the archeological and radiometric data from many, many prehistoric sites unearthed on the North American continent, there are none older than 12,000 B.C.

As mentioned above, plate tectonics and thereby the positions of land masses play an integral role in the occurrence of glacial periods. Greenland and Antarctica, therefore, are important components of our latest ice age. Greenland is huge! It has an area of 2,175,600 square kilometers (840,000 sq. mi.) of which 84% is covered by glaciers. Antarctica is even larger consisting of 14 million square kilometers of which 13.72 million square kilometers are covered with glaciers! (There is only 27,000 square miles of ice on Alaska). The ice on Greenland is over 2 kilometers thick, as is the ice at the South Pole, yet, at the North Pole the ice is only about 3 meters thick! The reason for this is that there is no land mass at the North Pole, and thereby nothing to support the ice. This means that North Pole ice cannot get too massive. As it does it just sinks and melts back into water. Can you imagine if the Eurasian land mass became situated in the center of the artic circle and other factors were right for an ice age?

The effects that glacial periods have had on the topography of our continents have been profound. The last ice age of the Pleistocene epoch, designated as the *Wisconsinan* age in North America (termed the *Tyrrhenian* age in Europe), is the glacial period we are recovering from now. Three-fifths of all the ice in the world covered a large part of North America. Another one-fifth covered parts of Europe. It pushed glaciers that stopped 75 miles from Tennessee.

It is called the Wisconsinan age because its effects are most evident throughout Wisconsin. It was a huge ice mass that was not evenly distributed, exhibiting great irregularity in its line of maximum advance. According to John McPhee in his amazing work *Annals of the Former World*, the ice mass "failed to reach Pennsylvania but plunged deep into Ohio, Indiana, and Illinois. The ice sheets set up and started Niagara Falls. They moved the Ohio River. They dug up the Great Lakes." Where they stopped and receded from, they left evidence in the form of *terminal moraines*. These are huge masses of undifferentiated rock, sand, gravel, and clay. This is because glaciers weigh an incredible number of tons and as they slowly advance they crush, move, or pick up anything in their way.

It was the effects of glacial advances of the past that ignited great debates in Europe, on how large boulders and other "leftovers" came to be placed seemingly in the middle of nowhere. The reason for this is that the last glacial period and its effects were more contemporary. There were still glaciers in the mountains of Europe. James Hutton in *Theory of the Earth (1795)*, postulated that large solitary boulders and terminal moraines were a direct result of ancient extensions of alpine ice. He was ridiculed for this.

The idea that Noah's Flood fashioned the topography of the earth dominated and was rarely challenged. It was because there were still ice masses or glacial extensions high in the mountains that gave the curious a first hand way to study the reality of glaciation, that evidence to the contrary grew. By contrasting and comparing glaciers and their effects in the Alps with areas of grooved rock, polished rock, moraines, as well as the solitary boulder phenomenon, at lower altitudes where there were no glaciers, the theory of continental glaciation fell into place.

When the Helvetic Society met in Europe in 1837, Louis Agassiz, the president-elect, deviated from his planned speech in paleontology to deliver, at great length, the evidence and chronology of glacial history. During this discourse he carried things a bit far but what he said mainly was correct. Still it received a chilly reception.

Agassiz, an expert in fish fossils, set out to gather more evidence. He did just that and wrote his *Study of Glaciers* (1840). Still there were not many scientists that would accept it. Finally, the most prominent geologist of the age, Charles Lyell, read the study and pronounced himself a believer. By 1862, even the scientific elite at Cambridge and the Geological Society of London came to accept it after many went into the field themselves to compare and contrast the evidence.

It's not too difficult to figure what effect global warming might have on our world. As carbon dioxide levels rise thereby contributing to the greenhouse effect, the temperature of the earth rises, too. The huge amount of water tied up in glaciers in Greenland, the Antarctica, and on mountains throughout the world would slowly melt (as evidence indicates is occurring). If all that water ends up in the oceans, the sea levels would rise at least a hundred feet. This would mean that half the cities in the world would be under water!!

Humans have much at stake, obviously, in the curtailment of this phenomenon. Unfortunately, getting "near-sighted" *Homo sapiens* to consider what's down the road a generation or so is not within our operational framework...especially if they are invested in the capitalistic mechanism, that has much inertia, that is causing it. What it will take is disasters, with loss of property and life on a global scale, to convince most. This is a sad but true fact of our species.

Humans, Culture and Civilization

What started civilization and the early cultures? What made people "decide" to live together in one place as a group thereby setting in motion events that led to the Egyptian civilizations, the Roman civilizations, etc., through to today? It was stated in the overview that humans existed in nomadic bands of hunter-gatherers for thousands and thousands of years before the first agrarian or farm-based society started. Proto-humans, called *Homo erectus*, who were the first of our distant ancestors to migrate out of Africa, also foraged in mobile groups of hunter-gatherers long before that, and their forebearers did the same before them. What was it that caused our most recent ancestors to "settle down"? Did the leader of a particular group finally just say "I've had enough, I can't walk anymore, we're staying here"? What does the archeological evidence say about this and where on the globe does it show that we first "settled" and why?

The *Neolithic period*, or when humans made their living from food production rather than hunting and gathering, appears to have started in the Fertile Crescent in Southwest Asia about 10,500 years ago (8500 B.C.). It seems to have centered around three particular types of naturally occurring wild cereals that had unusually high protein contents. They are *emmer wheat, einkorn wheat*, and *barley*. Soon after, the domestication of five other "founder crops" occurred in the same area cementing the emergence of the Neolithic period. The others were the *pulse lentil, pea, chickpea*, and *bitter vetch*; and the fiber crop *flax*. Of these eight crops only two had a natural range outside the southwest Asia/Anatolia area, and those were flax and barley. Soon after this the olive was "domesticated" as were the first animals, the sheep and goat (along with the dog).

So this takes us back to the question, what precipitated the change? The range of elevation in the Fertile Crescent contributed greatly to the "conversion" from hunter-gatherer to farmer. Within a relatively short distance, from the Dead Sea to the 18,000 foot mountains in Iran, these eight founder crops grew naturally. Since the altitudes varied there, there was an automatically staggered harvest season with plants at higher elevations producing seeds later than plants at lower elevation.

The fact that this area has mild wet winters followed by long, hot, dry summers, helped early farmers, also.

Many of the aforementioned eight crops are "selfers" which means you don't have to do anything to make them grow, just know where they are located, and be willing to harvest.

So, one can imagine hunter-gatherers frequenting this area every year, for the bountiful food available. All they had to do was figure out a way to harvest enough and store it to last them through the rainy season, the early part of the next growing season, and then repeat the sequence again. Then they could become sedentary. They appeared to have solved storage problems with the advent of *pottery* around this time. Through the domestication of goats and sheep, refining crop storage, making harvested crops into less perishable foodstuffs, and augmenting all this with hunting, their "society" gained momentum. As time went on and their abilities improved they were able to generate a great *food surplus* (which meant that every person in their group did not have to work constantly either hunting or in some agrarian capacity). This became the cornerstone of civilization.

Once food surpluses were generated, another key ingredient to civilizations was possible. This was *specialization*. No longer did all energies of the group need to be focused and expended for the mere survival of the group. Now there could be warriors to protect and defend (and later to engage and conquer), political leaders to address wants, concerns, and needs of the group, and religious leaders to satisfy the spiritual requirements of the group. Other specializations occurred also, as numerous as one can imagine.

The other very important effect that the transition from nomadic hunting and gathering to sedentary agrarian living had, was on the birth rate and hence on the population. Hunting and gathering allowed a couple to have a newborn every four or five years. Breast feeding needed to be extended after childbirth to several years due to inconsistent food collection and distribution inside the nomadic unit. The release of *prolactin* (lactational hormone) has an inhibitory effect on ovulation and therefore acted as a normal contraceptive until after the newborn matured to the point that they could eat what the troupe was eating. When farm based living started, especially since it was centered around cereals, the length of time before there was a switch from mother's milk to alternates (like cereals) was shortened dramatically hence removing the natural contraceptive nature of breastfeeding and thereby increasing the birth rate.

The rise in population of early "civilizations" necessitated specialization, especially in the form of "conflict-resolvers", that would arbitrate the inevitable disputes between the growing number of parties living in close proximity to one another. This is a direct result of the violation of the *Rule of 150* that had evolved in human hunter-gatherer societies in the previous thousands and thousands of years. As mentioned in earlier sections, the human brain's ability to think in terms of incredibly huge numbers or in great expanses of time is restricted. So, too, is the human brain restricted in its ability to function in groups greater than 150 without conflict. S. L. Washburn, a prominent evolutionary biologist wrote:

> "Most human evolution took place before the advent of agriculture when men lived in small groups, on a face-to-face basis. As a result human biology has evolved as an adaptive mechanism to conditions that have largely ceased to exist. Man evolved to feel strongly about a few people, short distances, and relatively brief intervals of time; and these are still the dimensions of life that are important to him."

Primates have the largest brain to body ratios of any mammals. The *neocortex*, which deals with complex thought and reasoning, is huge in primates as compared to other mammals. Social order and the complexities of larger social groups was a natural selective pressure on the evolving neocortex. Primates are social animals and there is a direct correlation between neocortex size and the complexity of social groups all primates live in.

Based on this, and in studying the various other primate species, their interactions, and the size of their social groups in the wild, it can be extrapolated that human social groups work best if numbers in a group don't rise too far above 150. British Anthropologist Robert Dunbar states: "The figure of 150 seems to represent the maximum number of individuals with whom we can have a genuinely social relationship, the kind of relationship that goes with knowing who they are and how they relate to us."

It seems to fit, too, because of 21 different hunter gatherer "societies", that have abundant archeological evidence which to analyze, none were greater than 150. This data does not just come from one part of one continent, either. Components of this study are as disparate as nomadic groups in Australia (the Walbiri), New Guinea (the Tauade), and Greenland (the Ammassalik). There's even a religious group, presently, called the Hutterites (related religiously to the Amish), in Europe and America, that splits off to a new colony every time the number 150 is exceeded. When things get larger than 150, people become strangers to one another, they are known to say.

So the fact that specialization, in the footsteps of food surplus generation by early agrarian societies, paved the way for larger populations and hence true civilizations, is obvious. But it was only because we were able to override the Law of 150, by that specialization (creation of conflict managers i.e. judges, politicians, policemen, and religious figures). After that problem was solved, specialization in other forms, especially as tool/weapon makers and soldiers, led to the Great civilizations.

Back to the original point, I don't want to sound like the Fertile Crescent was, by being the first place to start farming and with it a society that revolved around the sedentary population, the only area on the planet to independently start a "civilization". There were eight to ten other areas that did so, albeit at a later time and with different "crops". For example, China domesticated rice and millet as well as the pig and the silkworm, only about a thousand years later. Preludes to the Mayas, Aztecs, and Incans did it in the New World around 3500 B.C. with corn, beans, squash, and manioc as well as the turkey, llama, and guinea pig.

Inhabitants of the Fertile Crescent, though, because of the best species available to them, enjoyed quite a head start. But, it wasn't like you can keep a thing like that a secret. The ideas of agriculture, in no time, along with pottery, branched out to the northwest into Europe and east into Asia.

The wheats and rye (the main components of the founder crops) did well at many latitudes provided there was an ample growing season and the Eurasian continent, being huge in an east-west direction, was perfect for their dissemination (same growing latitude). Also, each new area in which agriculture was tried, contributed, it seemed, a new crop or animal to be domesticated to augment the founder crops. The poppy and oats in Western Europe, sesame seed and eggplant as well as humped cattle in what is now part of India, and in Egypt the sycamore fig, chufa seed as well as the donkey and the cat, all were domesticated *following the arrival of the founder crops*. By 1 A.D. cereals of the Fertile Crescent were growing from the British Isles to the coast of Japan and some of the worlds great early civilizations had already come and gone.

Besides the founder crops and especially the two wheats and rye, the next most important ingredient to get a pre-civilization really rolling is domesticating large animals as beasts of burden. There are 148 large *herbivorous* mammals in the world, that could've worked in this capacity yet only 14 ever became domesticated (large *carnivores* couldn't work due to their energy demands).

It's pretty obvious but the reasons why most couldn't be domesticated was because of one of four reasons or a combination of the four. These are 1) nasty dispositions i.e., grizzly bears, the hippopotamus, zebras, or the rhino 2) Tendencies to panic i.e., high strung gazelle and elk 3) inability to breed in captivity i.e., vicuna 4) growth rates and delayed maturity as with gorillas and elephants (elephants take 15 years to mature to a point where they can be used so they are just captured as adults).

The Eurasian continent had the most species (of the 14) and what turned out to be the best species, thereby giving them another advantage over emerging civilizations in the America's and Australia/New Guinea. The way in which large animals enriched our lives was four-fold. They furnished food (meat, milk, etc.), fertilizer, aided transportation, and pulled plows. Important animals to early Eurasian civilizations were the cow, horse, water buffalo, and Bali cow (and yak/cow hybrids).

The one large domesticated species that conferred the greatest advantage to early human civilizations, was the horse. First domesticated on the steppes north of the Black Sea around 4000 B.C., horses not only helped in the four ways just stated, but they also enabled a new concept in human-human interaction: *mobile warfare*. They made possible the rapid transport of large numbers of soldiers over great distances, an attack by surprise, and a quick exit from the campaign before a superior defending force could react.

Horses absolutely revolutionized warfare when they were first used by Indo-European speaking warriors that implemented them in their westward expansion out of the areas today known as the Ukraine. Subsequent spread of the use of horses and improvements of saddles and stirrups allowed the Huns and successive groups of other peoples to engage and terrorize the Roman empire and later civilizations to the point that they dominated much of Asia and Russia in the 13th and 14th centuries.

When Pizarro subdued the Incans at Cajamarca, in the Peruvian highlands, in November of 1532, he did so with the use of horses. Using 62 soldiers mounted on horseback and another 106 foot soldiers, he killed over 7,000 Incan warriors and dispersed a force of over 60,000 that had assembled to meet them. The Incans had never seen horses before and on a given signal, the cavalry burst forth from hiding. Equipped with rattles as noise-makers to frighten the natives, they "set upon them with swords and cut them to pieces". This is discussed at length in Jared Diamond's Pulitzer Prize winning book, *Guns, Germs, and Steel.* In this great work, he underscores, adroitly, the great advantage that horses gave those who possessed them over those who had none.

So what really was it that separated early villages, societies, and cultures from civilizations? What exactly constitutes a civilization? Civilization is defined as "an advanced state of human society, in which a high level of culture, science, industry, and government has been reached." If culture is "the sum total of ways of living built up by a group of human beings and transmitted from one generation to another", then cultures, through advances in science, industry, and methods of governing, can attain civilization status. This is exactly how the earliest known civilizations came to be.

Starting as villages in a part of the Fertile Crescent known as Mesopotamia (literally "between the rivers" of the Tigres and Euphrates) the *Sumerian* culture was transformed through advances in science, industry, and government, to status as the first true civilization starting around 3500 B.C.

The city-states of *Ur, Lagash, Nippur,* and *Kish* built up around the oldest city of *Eridu*. So it took inhabitants of this area from their start in 8500 B.C., 5000 years to reach the point where they were somewhat united under a common government and culture, to where they would be designated as a true civilization. It really wasn't until 2350 B.C., though, that Sumerian city-states were truly united. This was under *Sargon* of *Akkad* who was the first "great" ruler in the world.

Their civilization was conquered by the Babylonians in 2000 B.C. and boasted, nearly 250 years later, the great ruler named Hammurabi (known as Amraphel in the Old Testament). Hammurabi made advances in governing by publishing his code of laws called *Hammurabi's Code*. This is thought to be the first such code in history. The Sumerian and Babylonian civilizations were the first to use a written language (called *cuneiform*) and their culture was steeped in a polytheistic religion that believed in *animism, anthropomorphism,* and *life after death*. They invented a math system based on the number 60 and made advances in geometry and algebra. They started a system of weights and measurements and mapped many constellations. The Sumerians are also credited with inventing the wheel around 3700 B.C.

The *Egyptian* civilization started around 3100 B.C. when *Menes* unified populations of upper and lower Egypt. The civilization lasted till 332 B.C. and thereby was significant as a civilization almost 2800 years. The Egyptians had their own system of writing called *hieroglyphics* and their own polytheistic religion in which the *Pharoah* (leader) was able to communicate with the gods. They built the amazing pyramids as early as 2600 B.C. and contributed their own advances in science, medicine, and astronomy.

The Chinese (3000 B.C.), the Minoan (2500 B.C.), Greek (2000 B.C.), and Etruscan (1000 B.C.) civilizations rose also in the Eurasian theater, in different areas.

In the New World the Mayan civilization arose (1800 B.C.) followed much later by the Aztecs and Incas.

Rome, India, the Hebrews, the Vikings, and Japan followed.

Then as populations grew and expanded to every corner of the globe there were cultures and civilizations nearly everywhere.

So what was it then that enabled European civilizations to come to dominate the globe in the 1500's and beyond, setting in motion the extensive holdings that came to be possessed by Great Britain in the 1800's ("the sun never sets on British soil")?

European societies had the benefit of the head start afforded to them by the "founder crops" and after that it was a technological race that Europeans came to win. The reason for this is pretty simple. The Chinese, who started a bit later, were unified in relative peace under the *Dynasties*. The Europeans were subdivided into differing cultures in relative close proximity to one another and they had a difficult time getting along with one another. This led to conflicts which led to a technological race to see which culture could come to dominate the others. Most technological advances, that helped one society over another, were because of improved abilities to kill. Better steel, better catapults, Greek-fire for dominating the seas, better ships to control the Mediterranean, better tactics, better logistics enabling control of more resources and more resources. They always fought and got better at it because they started earlier and couldn't get along with each other. The adage "might makes right" came to dominate and after that it was a quest for resources.

Resources and their discovery, their integration into the expansionistic policies, and the quest for more resources and the riches they brought with them, historically, drove the machinery of war. Resource management, now, due to the world population being so great, has equal but different importance in the 21st century. The relatively peaceful time we live in does not preclude the need for resources. The need has become more economic than purely militaristic. Militaristic technological advantages appear to drive and protect economic expansionism, though.

Resources, and their procurement, were the centerpiece of all past civilizations. They continue to be the focus of present day civilizations, also. As humankind's total population increases worldwide, our earth civilization revisits today, and will revisit in the future, resource management problems that past civilizations wrestled with.

In antiquity, resources necessary for the perpetuation of a civilization, were simple yet undeniable. A water supply that was clean and reliable was paramount.

Building supplies, most often forest-related, were also critical. Soil that was fertile was essential (and there had to be enough arable land to allow expansion of the population). Animals in the wild, that were either domestication-friendly or abundant enough, so they could augment food supplies, were desired. This latter point encompassed fish and the fishing industry that drove many early civilizations and cultures.

Nowadays, due to mass production and refrigeration, not to mention transportation improvements that have ushered in the term "globalization", resource management, though just as critical as in the past, has become much different. Advances in science and technology have made our lives so much easier. We no longer have to spend all our waking hours attempting to secure enough food to survive. Yet, with this improvement in providing for our populace comes a world population that keeps growing and growing and growing.

So the question becomes, "do we have the resources to accommodate the population of the world in 10 years, in 20 years, etc.?" What other side effects does a burgeoning population, and mankind's attempts to provide for this population, have on the world's ecology? What can history tell us about resource management?

It's amazing that in the 5500 years since the earliest true civilization of the Sumerians, we have seen so many civilizations come and go. In so many cases, the demise of a civilization occurred, not so much because of poor resource management, but rather through the increased militaristic might of an invader and their expansionistic policies. From these, we can learn only to be vigilant in the face of threat, by maintaining a sound defense.

What of the civilizations, though, that disappeared from the face of the earth only through the mismanagement of their resources? Of these are there any parallels between their situation and the situation the world is either in today or fast approaching?

Societies of the Anasazi in what is now New Mexico, of Polynesians that came to inhabit Easter Island in the South Pacific, of the Mayans in Central America, and of the Viking settlements on Greenland, are four "civilizations" that doomed themselves due to poor resource management. There were other factors involved, to a degree, in each of these situations, but the most important factor by far was underestimation of environmental fragility and exhaustion of resources that were critical to survival.

Deforestation followed by soil erosion seems to be the biggest blunder in each of these groups.

On Easter Island especially, an island 1,300 miles from its closest neighbor, the lack of tall trees was so complete that archeologists figured, at the time of initial settlement, that there never had been any big trees. Just the small *toromiro tree* (only 7 feet) had any size at all among the 47 plants that grew on Easter in the 20th century. It is well known that the civilization that inhabited this remote island, in the centuries after initially settling there, built some incredible statues. Everyone has seen pictures. There were 393 of these monoliths on platforms erected at various places around the 66 square mile island. There were 97 others completed, that appeared to be in transport to unknown destinations on the island, when abandoned. Amazingly, there were over 300 more statues in different stages of completion and of different sizes, in the rock quarry that all had come from. How, the heck, could these chunks of stone, some weighing as much as 70 tons, be transported around the island, without large trees to roll them on and form fulcrums and pulleys with? This mystery led to books about aliens, being marooned and finally being rescued, and other supernatural phenomenon.

The puzzle was cleared up by *palynologists* (scientists that study plant pollen). Pollen, from all botanical sources living in a given area, gets deposited around water sources, over hundreds and hundreds of years, in layers with vegetation that dies, and other sediments. The most recent hundred year deposits are on top with successive years stratified or layered underneath. By *stratigraphic* analysis of pollen, it was surmised that a large palm tree, very similar to the Chilean wine palm, only larger, had grown throughout the island in the past. Some "fossilized" palm nuts were also found in a cave by archeologists, further corroborating the palm's existence on the island. Further evidence in the form of palm trunk casts from ancient lava flows was also discovered. All this led to the obvious conclusion that there were lots and lots of these huge palms when the initial settlement occurred.

Further evidence, that the staple of the early islanders, as deduced from the excavation of garbage *middens*, was a mix of dolphins and tuna, came to light. This fit with the existence of the large palms. The reason why this fits, is because the only way to obtain these bigger sea creatures was by going out into unprotected seas to harpoon them. They didn't frequent the shore areas. The only way the natives could have done this was to make use of the very large palms to make large, seaworthy canoes.

Also, the time when the last palm disappeared coincides with when the islanders stopped eating these larger sea creatures (as determined from stratigraphic analysis of their garbage middens, finding these dolphin and tuna bones at the earliest times, continuing to find them for 5 centuries, and then at subsequent upper layers not finding them…ever again).

Radiometric dating of charcoal shows the last of the big trees being burned around 1500 A.D., or so. So all the evidence says the inhabitants deforested their island and left themselves without a means to secure their primary food. This begs the question, as postulated by Jared Diamond, in his new, compelling book, *Collapse: How Societies Choose to Fail or Succeed*, "what were they thinking and saying to each other when they were cutting down the last of the large palms?"

After this "boo-boo", Easter Islanders relied on domesticated chickens, food they could grow on the less and less fertile soil, food they could garner from the seashore, and rats.

Rats, in fact, were not naturally occurring on the island, having stowed away in either the earliest colonizing vessels or on vessels that infrequently came afterward, to trade, from other islands…they, in turn, became part of the deforestation problem because they would gnaw on the palm nuts that could have led to more large palms, thereby preventing them from germinating. The fossilized palm nuts, mentioned previously, had gnawing marks on them, indicating the pestilence.

Garbage middens showed also the tendency to rely on rats as a staple, after the 1500's, when things got bleak. When things worsened, they survived by resorting to *cannibalism*. From a population estimated to be 15,000–30,000 at its zenith in the 1400's, it deteriorated to only about 3,000 or so, when in 1836, to add insult to injury, about 1,500 inhabitants were taken away as slaves by Peruvian slave-traders.

So what could've been done differently? It's very difficult for humans, when there is prosperity, to confront the fact that resources don't last forever. It's easy for us, as 21st century humans, in retrospect, to wonder, in amazement, "what were they thinking?" I'm sure, though, that in 100 years from now, there will be people on our planet, with the benefit of hindsight, that will be wondering in amazement, what we were thinking when we were continuing our over reliance on fossil fuels. Or, they will wonder why we couldn't see the warning signs of global warming. Or why we just sat back and let our world population grow ever larger, beyond the capacity of the farmers to feed all of us. Or why we continued to expand into ever farther reaching areas, compromising the diversity of our planet, sending it further and further from the equilibrium that took so long to establish. An equilibrium that ensured that most species would have a chance for survival, not just the one we've come to think of as the chosen species.

The reason why societies fail to see the writing on the wall is actually pretty simple. Power resides in the hands of a few. The few are heavily vested in the status quo. Milton and Rose Friedman co-authored *Tyranny of the Status Quo* and in this work they indicated how difficult it is for our American society, with all its politico-socio-economic inertia (in the form of our bureaucracy), to change course, even just a little bit. With two, four, and six year terms for our politicians and civil servants, mega-corporations concerned only with the bottom line and all their money involved in lobbying efforts, not to mention constituencies that are concerned mostly with the here and now, the system can not look further than ten years down the road, maximum. This is the way it is in our "civilization" and this was probably the way it was on Easter Island. In fact, their ability to look ahead was obviously much shorter if the last tree of value was cut down without addressing the reason why that might not be such a good idea.

The Vikings, when they first settled in Greenland, similarly paved the way for their own demise through deforestation. Loss of trees led to soil erosion in their fragile environment, which led to a difficulty maintaining necessary livestock levels. Couple this flawed approach with an inability to adapt to their environment completely and it is surprising that they lasted as long as they did (over 400 years).

In Norway, the Vikings were used to eating beef and they tried to force this Euro-centric attitude on an environment that resisted. With the seas all around them, with unbelievable numbers of fish waiting to be harvested, the Vikings shunned this food-source, even when starving.

Instead of concentrating on iron tools and weapons that would help them survive, the Vikings had religious statues and stained glass windows shipped in from Norway in order to adorn their churches and cathedrals.

They were belligerent with any other societies they came in contact with (earlier the Amerindians in New Foundland, or "Vinland", and later the *Inuits* or Eskimos) thereby stifling any "adaptive strategy" flow of information that would most certainly have helped them. The Vikings looked upon the Inuits as pagans, and with arrogance they attempted to maintain their Norwegian way of life. It ended up costing them dearly.

Islands of the South Pacific named Pitcairn and Henderson had "civilizations" that endured similar extinction for similar reasons; fragile environments that buckled under population pressure from humans unable to see signs of problems until it was too late.

The Chaco Anasazi suffered a similar fate in their fragile, resource limited area, in what is now northwestern New Mexico (Chaco canyon). This "civilization" is thought to have made the conversion from hunter-gatherer to a farm-based sedentary society, along with other Amerindian cultures, around 1 A.D. There is still some debate on this due to the fact that evidence indicates there was still a bit of movement during the non-growing season. Later, as buildings in villages were erected, the evidence became clear that these societies were, indeed, truly sedentary. Solid proof, that by 600 A.D. the Anasazi were going strong, is irrefutable.

For five centuries they lived and flourished until some time in the late 1100's they disappeared. Toward the end, things became so bleak for them, that they too, like the inhabitants of Easter Island, resorted to cannibalism in order to survive.

The Anasazi (from the Navajo language meaning "ancient ones") were quite impressive in their ability to understand their environment. They built dams to store rainwater and control water flows coming from the tops of cliffs rimming their canyon. They built large buildings up to six stories high, with over 500 rooms, using logs 16 feet long and weighing 700 pounds, as roof supports. They planted crops in areas close to the water table (they relied on the "Mexican trinity" of corn, beans, and squash). They were a complexly organized society and regionally integrated via trade with other cultures 30 to 100 miles away. So what the heck, happened? If you went to the Chaco canyon area today you would not believe that it had been home to a civilization of over 5,000 people, its "high water mark", just after 1000 A.D. There is almost nothing there, no trees, just brush, dirt, and rock. What was it that attracted the early inhabitants that first settled there before 600 A.D.? How can we know what it was like?

Well, when you want to know something about a civilization that lived in the past, just as if you want to learn about a group of people living in the present, you look at their garbage. As discussed previously, archeologists studied the stratified remains of garbage middens on Easter Island, to figure out what past diets were like, how they changed, what was valuable to them, and what was worthless. Similarly, in Chaco canyon, garbage middens yielded clues as to what things were like but the most valuable tool, archeologists used, was from an unlikely source.

In the late 1840's, prospectors heading for the California gold rush by way of the Nevada desert, happened to notice, glistening in the sun, a rock-candy like substance, at the edge of an incline of rock. Being hungry, one of them tried tasting it and was pleased to find that it was sweet. The others joined the first in licking and eating the new-found "candy" until they began to feel nauseous. When examined closer the "candy" was found to be hardened balls of deposits, gatherings, and excretions of some animal. The animal, it was later determined, was the *packrat*. What the prospectors had been consuming was a *packrat midden*; in essence dried rat urine laced with rat feces and rat garbage!

It turns out that packrats live all over the place in this dry hot climate. They form a nest in a safe area and anything within thirty feet or so of the nest, will become incorporated in the nest. Leaves, pine needles, wood chips, or whatever will become part of the midden. The packrat releases its urine and feces right in the midden. It raises its family in the midden. Packrats actually pass on middens to their progeny. After a few decades, for some reason, these animals abandon their home and move on to start a new one elsewhere. The crystallized urine that has accumulated and accumulated actually prevents the midden, in the hot climate, from deteriorating.

A midden, to the glee of archeologists, is a time capsule of the different plants that occupied the area, at the time of its formation. Unbelievably, as determined through radiometric dating, packrat middens can survive, if protected from the rain and other elements, for up to 40,000 years. It was by collecting bunches and bunches of these "time capsules" and radiometrically dating them, that they were able to solve a mystery.

They analyzed all the middens, back through time and were able to show that the area was forested as a *pinyon/juniper* woodland before they arrived and through their early settlement, all the way till about 1000 A.D. At this point and thereafter, packrat middens lacked pinyon pine needles and juniper needles. This indicates deforestation in their canyon was complete around this time. It was this exhaustion of a critical and finite resource that set in motion the decline of this society. After this, inhabitants had to venture some fifty miles, in order to secure the wood that was needed by the populace.

This was made all the more difficult since there were no beasts of burden at the disposal of the Anasazi and therefore large logs were carried by bands of workers all that way. Oddly enough, despite this problem, the population appeared to have continued to increase, due to better than normal rainfall and therefore greater food surpluses. It was soon after that the last of their construction activity occurred (1117 A.D.). This coincided with a series of droughts from 1090 to 1130 A.D. that appeared to be the final blow to this society. The droughts, the need to send so many men to get wood so far away, a decrease in natural game, and emerging conflicts from neighbors meant the end around 1175 A.D. The last inhabitant left the canyon never to return.

The Anasazi were impressive in understanding their environment, doing better than the Vikings in Greenland, yet they too, dropped the ball in not comprehending the deforestation problem and its effects, the stress a rise in population would have, and how it would all play out if crop yields were low. They lasted about six centuries, which is almost three times that of the United States, so that isn't too bad.

It should be mentioned here that another tool that archeologists have at their disposal is that of *dendrochronology*. Dates above, stating when building ceased as 1117 A.D. were exact, due to this science. It involves reading tree rings and analyzing growth in wetter abundant seasons versus drier, poorer seasons (and comparing the sequence of ring size in older trees, and continuing backward). The fact that trees grown and used for construction in this area commonly lived 150 years prior to being cut down, is of great importance to *dendrochronologists*. Also, the fact that in such a dry climate, the wood used in construction still standing, was nicely preserved (all the way back to the earliest times), was critical. This is just another tool in our examination of the failings of a particular civilization thereby better illustrating the inherent inability of humans to anticipate problems too far down the road.

By way of selective pressure on our brains, in the expanse of time during which we as a species lived only day to day, we've come to be deficient in the capacity to "forward think". As mentioned previously, we can at best, think 10 years down the line, environmentally. At worst, we will as our failed ancestors, merely react with band-aid solutions, and just carry on. This is what we are programmed for. So what are the signs of trouble to our own civilization today? What does the populace of the world need to fear and remedy, if we are to survive?

Just as with the Anasazi, the Greenland Norse, or the Easter Islanders, we live in an ecosphere that is much more fragile than we think. Industrialization that occurred in the late 1800's, gathered steam throughout the 20th century, and continues unabated today. The population of the earth is skyrocketing. It's over 6 billion now and predicted, at present growth rates, to easily surpass 12 billion by 2050. Our dependence on fossil fuels, their oxidation, and release of waste products of all kinds, into our atmosphere (especially carbon dioxide) and their effects, coupled with deforestation, is offering us vital warning signs of future mayhem.

The last critical frontier to all this, where the line must be drawn but as yet has not, is the area referred to as *Amazonia*. The Amazon rain forest is more than 11 times the size of Texas. It is home to one-third of the world's species. It has been referred to, by some, as the Earth's lung. Its capacity to absorb carbon dioxide, through photosynthesis, and release oxygen, is immense. The problem, though, is that it is being destroyed at an alarming rate. There were 600 forest clearing fires per day several years ago, attempts by inhabitants to clear patches to create cropland to feed increasing populations. The Brazilian government has plans to pave 2,100 miles of additional roads into the forest, in the near future that could lead to an additional 70,000 square miles being deforested in the next 30 years.

The capacity of the Amazon rain forest at present, to absorb more carbon dioxide than it releases, is barely being maintained. With more frequent and severe El Nino weather patterns predicted in the near future, conditions that favor a drying out of the large rain forest, things appear to be getting worse. The Intergovernmental Panel on Climate Change, a United Nations organized science network, warned recently that the remaining Amazon "is threatened by the combination of human disturbance, increases in fire frequency and scale, and decreased precipitation from *evapotranspiration* loss (dry forests release less water back into the atmosphere), global warming, and El Nino."

ICESat, a *NASA* satellite originally put aloft to monitor Antarctic ice mass and melting polar ice caps, has been redeployed over Amazonia, recently, to acquire data on how much wood is in the forest. *LBA* (Large-scale Biosphere-atmosphere Experiment in Amazonia), a Brazilian-U.S.-European scientific team, comprised of over 1,700 researchers from 200 universities, has been collaborating to collect and analyze data from the huge region. Other scientists are conducting still more experiments in both the Amazon and in far removed locales on the globe to measure changes to our ecosphere. The data being compiled is impressive and irrefutable.

But will we listen to the warning? If we don't curtail our emissions and we continue to pretend there's no problem, we run the risk of getting further and further in trouble. We are throwing the world further and further out of balance biochemically.

Of Aliens and Alien Species

SETI, or the search for extra-terrestrial intelligence, to be discussed toward the end of this section, is a rather ironic human concept since after considering the silly things humans have done on this planet, it may appear that there is little intelligent life here on earth.

The established equilibrium between continental species, regarding both plants and animals, that was entrenched over millions of years, is by nature, a tenuous one. Overpopulation by humans and the species grown to feed them, and all the byproducts and waste products emanating from them, has greatly disturbed this equilibrium on every continent, to some degree. This is evidenced by the constant disappearance of species from the face of the earth. Some causes for this were discussed to varying degrees in the CHNOPS section.

The term "equilibrium" can be defined as "the condition in which no detectable change occurs in the state of a system as long as its surroundings are unaltered".

Different things affecting a biological system, as it pertains to a particular plant or animal, ultimately leads to either a greater ability to survive, a lesser ability to survive, or lack of any effect. For example, over harvesting Atlantic cod by humans obviously is detrimental to the cod species. The cod species is also important to predatory shark and barracuda species that rely on it for food. Decreases in cod thereby decrease these species numbers or causes them to turn to other species. This results in a disruption in the general equilibrium with unpredictable results.

Releasing pollution constantly into a tributary obviously disrupts the natural equilibrium of the river and hence negatively affects species that reside therein. One could go on and on discussing species after species in niche after niche that humans have disrupted beyond recovery, due to ignorance or disregard. The topic of this section, though, concerns an alternative way, again through ignorance and/or arrogance, that humans have disturbed tenuous equilibriums all over the planet.

Alien species introduction, or introduction of a species of plant or animal that is naturally occurring on one continent or even part of a continent, into another continent or part of a continent (where they are not naturally occurring) can be a recipe for trouble. Introduce into the equilibrium established between many species, a competing species (that competes better for the same food and nesting sites), and you may have a species that is greatly affected by this introduction, possibly to the point of extinction. A good example of this point is the sad story of the Starling (*Sturnus vulgaris*).

The Starling was introduced into North America in 1890 by an English fellow who thought that all bird species mentioned in the works of Shakespeare should also live in the New World. He transported 100 of them across the Atlantic by boat and released them in Central Park in New York City. From this initial seeding the Starling population in the U.S. alone is thought today, to be well over 200 million. Oh, by the way, thanks, dufus! They have disrupted the equilibriums of all "cavity-nesting" birds with the main damage done to woodpecker and bluebird populations. They have been so successful because in this new world of theirs the normal predators, that keep them in check in Europe, are not present. They are the only bird species that I know of that is killed by humans, on a large scale, to control their populations. This is an example of an *intended introduction*.

Additional examples of problems associated with intended introductions of alien species involve the rabbit and fox relocation into Australia, in the 19th century. Both were introduced at about the same time, though not together by the same group of people. The reason for these introductions, besides ignorance, was to attempt to make Australia like England. Englanders that had come to live in Australia found the naturally occurring animal species there foreign and longed for species that made them feel at home. What, after all, is more British than fox-hunting? And naturally, if you were going to bring a bunch of foxes to a new environment you should bring them what it was thought they would want to eat, namely rabbits. At least that's how the thinking process was thought to have occurred.

Well, domesticated rabbits were released in this little venture, and failed to survive. So three more times, white, tame rabbits were released, only to die off. Finally, to everyone's joy at the time (and everyone's horror 100 years later) wild Spanish rabbits were introduced and took firm hold. They proceeded to do what rabbits do so well, they multiplied like, well, rabbits. Foxes were similarly successful. The amount of time that passed, before people realized that a blunder had occurred, wasn't too long. Huge rabbit populations led to soil erosion because the rabbits would eat the native grasses down to nothing and the grasses would then die. The soil, exposed since the vegetation was suddenly absent, would dry out more easily from direct sunlight and heat. The Australian continent being an extremely fragile environment on par, amazingly, with Greenland, couldn't regrow grasses fast enough. Without the vegetation's roots to hold the soil, the soil would just blow away in areas, or be washed away by rain.

The rabbits also would dine on crops that farmers would grow creating a severe economic impact. This included fodder intended for domestic livestock; flora normally eaten by native herbivorous mammals.

Foxes were no help in keeping the rabbits in check. They had other prey that were much easier to catch. This is an entire continent where all species of animals evolved for eons without any predators like the fox. Without an ingrained instinct to evade the fox, several species of small mammals went extinct in just 50 years after introduction. Foxes could literally walk up to these mammals and there would be little attempt to flee!

This is the same reason why within 100 years after Polynesians settled on Easter Island, flightless birds all but ceased to exist on the main island. The same reasons why the Maori of New Zealand were able to drive the large flightless bird species, called Moa, to extinction.

If either flightless bird species had evolved in the presence of predators that they could have evaded by flying, they would've retained the ability to fly. Then there wouldn't have been the easy extinction of these species.

The fox species aside, the main problem for Australians became the ubiquitous rabbit. At the time of introduction of the rabbit and fox, several species of bird were also introduced: the common House Sparrow, the Starling, the Blackbird, the Song Thrush, the Tree Sparrow, the Goldfinch, and the Greenfinch. The Starling became a minor problem and the House Sparrow also became widespread but the other species only survived in small areas. Why did the rabbit become such a problem? Once again, aside from its adaptability, it was due to the lack of natural predators.

Subsequently, man, with his great knowledge, decided to correct the problem by introducing a virus lethal to rabbits, called *myxoma virus* (a pox virus that causes *myxomatosis*). It was originally very successful, killing over 90% of the rabbits. Natural selection, though, working through mutations, allowed the remainder to become resistant and the population rebounded to previous levels. There is now another program under way using a different virus (*calcivirus*) that will most likely end up the same way.

Problems of intended introduction of non-native species involving plants can be even more serious than animals...especially if considered in the context of economic impact.

Rubber vine is an ornamental shrub naturally occurring on the island of Madagascar. It was thought, when introduced into a few gardens in Australia around Queensland, that it could be contained, and would add beauty and luxury to each garden. Unfortunately, its seed pods are rather adroit at disseminating their contents and this plant has also become a major problem to Australia. It is particularly suited to its new environment and without natural "growth restrictors" (natural predators). It grows huge vine masses which snuff other plants out by superior competition for light, water, and soil nutrients. Unfortunately, it is also toxic, so grazing animals have to be kept away from it.

An American problem plant, *Mile-a-Minute Weed*, was a naturally occurring vine indigenous to Asia. It was brought over to a plant nursery in York County, Pennsylvania, in the late 1930's. Although it is on record of occurring previously in two other locations in the U.S. it appears that this Pennsylvania site is the first area from which a self-propagating wild population arose in the States. It is an aggressive grower and has spread to occupy a 300 mile radius in and around York County.

Kudzu (a member of the pea family *Fabaceae*) is a big problem in the southeastern United States. It was first introduced to the Americas at the Centennial Exposition in Philadelphia in 1876. It was promoted as a forage crop and ornamental plant. It was actually planted by the government to reduce soil erosion in some areas but started to over compete and force out native desired species in the first half of the 20th century. It was placed on the pest weed list in 1953 and removed from the governments list of suitable cover plants. Its vines can reach 100 feet in length and as many as thirty vines can emanate from a single root system.

Other species of plant introduced purposefully only later to be recognized as a major problem in the continental United States are: *Fiveleaf akebia, Porcelainberry, Oriental bittersweet, Climbing euonymus, English ivy, Japanese honeysuckle,* and a couple of exotic *Wisterias*.

Hawaii has an unusually egregious problem with native species being dominated by mostly intended introduction of non-native species (although a very significant percentage were unintended). The Islands have had over 4,500 plant species that are not part of the normal flora, introduced in the last 200 years, for one reason or another. The main problem of ecological stress lies with 86 species, though. The most serious threat comes from the noxious weed *Miconia calvescens*. It has become so widespread that chemical eradication is no longer an option. The world is being combed at present for natural enemies of this plant in order to bring it under foot. Among the most promising remedies is a *Euselasia* species of caterpillar. This caterpillar is a voracious eater of leafage with a preference for this plant, apparently. One can imagine a nightmare scenario if this approach is pursued. It's difficult to establish million-year equilibriums in a year.

Problems of *unintended introduction* of species are even more numerous and just as troublesome. One particularly interesting species is the *woody aster*, a plant. It is not naturally occurring in Wyoming. The seeds were inadvertently brought with herds of cattle and sheep and later in trucks and railroad cars transporting hay from the south. The odd thing about the woody aster is that it draws selenium from the soil when germinating (it's required for germination). Selenium is a toxic, non-metallic element. It happens to be in high levels in some volcanic ash. Eons ago volcanic activity in the area released ash that had high levels of selenium. This ash settled in the Wyoming area. The area was covered with water at the time and the selenium became deposited in the lime muds upon drying. After the waters had receded and plant life grew, native species did not extract the selenium. The woody aster changed all that. Once the selenium is absorbed during germination, it is changed chemically to a form that nearly all plants can inadvertently take up.

When grazing animals eat plants that have certain levels of selenium, they suffer from a malady called the "blind staggers". They lose their sight and just stumble around. This is because the selenium interferes with normal muscle-nerve function by way of a particular enzyme's degradation. People also are affected. They get "dishrag" heart and can also suffer from liver damage, kidney damage, birth defects, and sterility. Millions of acres have been rendered unfit due to this problem. All of this is because of an unintended non-native species introduction.

Another plant species causing problems not just in the United States but also worldwide is the aquatic plant *Eichornia crassipes* known by its common name *water hyacinth*. First introduced into the U.S. from South America at the World's Industrial and Cotton Centennial Exposition in Louisiana in 1885 it was apparently taken and released by a Florida visitor into the St. John's river and from there has spread to nearly every lake and river in the South, it seems. It causes problems for navigation into waterways due to its rapid growth. It also creates a great breeding area for certain insect pests. There have been programs to control the growth of this difficult plant for many years with limited success. It continues to be a major problem worldwide.

An animal species causing huge problems but introduced much more recently into the U.S. is the *zebra mussel* (*Dreissena polymorpha*). This small creature was first discovered in Lake St. Clair, a small lake between Lake Erie and Lake Huron, in 1988. Indigenous to Eastern Europe they have been causing problems in

Western Europe since first noticed after the Industrial revolution. The mode of access to the U.S. was thought to be by the bilge water of some trans-oceanic freighter. Free swimming nearly invisible larvae called *veligers* were apparently discharged in the ballast water and gained a foot-hold in the Great Lakes. Now they are at the top of the list of foreign species to be controlled.

The problem is, again, there are few predators to control them. These pests disrupt freshwater plant and animal equilibriums by filtering out and eating phytoplankton in the 15–40 micrometer size. They are amazing reproducers and voracious eaters. In addition to this they also attach, by way of strong *byssus* threads, to any hard surface thereby clogging water intake pipes, screens for drinking water facilities, industrial facilities, golf course irrigation pipes, cooling systems of boat engines, and boat hulls. A concerted effort is now underway in the U.S. and Canada to prevent their spread and control their numbers.

The *cane toad* (*Bufo marinus*) is another example of good intentions gone bad. Australia, for some reason, it was found, had no species of toad. They had frogs but no toad had evolved there.

Sugarcane had also been introduced there and as with other areas in the Caribbean and Hawaii, the *cane beetle* was a problem for farmers. It was thought that the cane toad would control the cane beetle, so 102 were shipped into Northern Australia in 1935. They were bred in captivity briefly and then released. They proceeded to have no effect on the cane beetle but over the next 70 years have come to be a great problem to the rest of the inhabitants of the warmer climes of Australia. The problem is, except for a brief period as juveniles, the toad is poisonous throughout its life. It has no natural predators and it reproduces in huge numbers. They have out-competed almost all frog species and even eat frogs! They kill or sicken many species of domestic animals that they have encounters with. The tadpole is poisonous, too and fish eating them die. The toads are so poisonous that snakes have been found dead with cane toads in their mouths! They get very large if allowed to keep eating, which they do. Some toads, as big as 25cm and 4 kilograms, have been reported.

The entire experience has turned into a nightmare for Australians but there is a glimmer of hope. Some bird and rodent populations have acquired a tendency to kill the toads and eat only the internal organs. They stay away from the poisonous parotid gland.

For now, large sums of money are spent by the Aussies, annually, to attempt to control the toads.

Guam, an island in the Pacific, in the Marianas, approximately 2000 kilometers east of the Philippines, is a particularly disturbing example of what effect an unintended species introduction can have. Several *brown tree snakes* (*Boiga irregularis*), apparently stowed away in military cargo from New Guinea after World War II, and set upon the island. By the early 1960's the effects were unmistakable. Nine of Guam's twelve naturally occurring forest bird species were extinct with the other three getting close. Half of their native lizard species are now extinct, and several of the indigenous species of bat have been negatively affected. A snake control program is now in effect but is difficult. Most of the focus regarding this pest is aimed at not allowing it to stow away to other islands such as the Hawaiian Islands. There have been several reports of snakes doing just that only to be subdued on arrival.

The literature overflows with example after example of good species, when relocated, going bad, and bad species even worse, for all the obvious reasons. The stories of the *carp, lampreys* and their effect on commercial fisheries of the Great Lakes, *Spotted Knapweed* and *Leafy Spurge* in Montana, the blights that destroyed *American Chestnut trees* and *American Elms*, transplanted blowflies and ticks, among others, would make you sick. This doesn't even start to talk about the dissemination of bacteria and viruses, and their associated epidemics in a variety of species, alluded to in the previous sections of this book.

It would be a mistake, though, to think that it is a problem every time non-native species introduction occurred. The very basis of species domestication, whether plant or animal, which truly is the cornerstone of civilization, depends on introduction of non-native species.

We have corn and rice grown all over the world. Corn came from South America and rice from Asia.

Potatoes, an integral part of Irish history, are not Irish at all, but also South American.

Apples, as American as hot dogs? As you probably know by now, apples are distinctly Western Asian (Kazakstan)!

Honey and honey bees were brought to America, they were not indigenous.

Horses came from north of the Baltic Sea and from there were taken everywhere by man, it seems (no, they did not naturally occur in the Americas...an evolutionary cousin, only 3 hands high, did inhabit the Americas at one point in the distant past, though).

The pig came from China and South America and the sheep and goat from Southwest Asia.

Tulips are such a part of Dutch history, yet they originated in what is present-day Turkey.

Species relocation and the experiments with domestication of both flora and fauna have regularly punctuated *Homo sapiens* history. We have, through our mistakes, erected safeguards to regulate occurrences. Whenever one travels across nearly any border in the world information is available as to the travails of assisting such trouble. Questions are routinely asked to ensure the lowest probability of dissemination. Customs inspectors are doing good work. This is a great example of time and money, being spent worldwide, via international cooperation, to halt an out of control, destructive human practice in its tracks.

The point that mustn't be lost, though, is that our ecosphere, prior to human overpopulation, established an equilibrium over millions and millions of years. When one species overgrows to the point that humans have, there will be an effect overall. When the dominant species undergoes technological advances in transportation that make unintended movement of species common, introduction of foreign species, though inadvertent, will occur. Therefore the tenuous equilibrium will be further affected. Can our ability to advance our species technologically bail us collectively out of this mess we are creating? Time will tell...hopefully it will not come in the form of leaving the planet to settle elsewhere in the universe.

If in the future it did come to that, what would be the likelihood of our finding other life in the universe? What would be the likelihood of just finding a planet suitable for humans to settle?

The universe is incredibly large. Estimates of its size are based on things like the *Doppler Shift, Cosmic Microwave Background Radiation,* and *fractals.* Depending on the source, one can secure estimates ranging from 27.4 billion light years (a light year being of course the distance light travels in a year) to infinity.

The likelihood that there is life elsewhere in the universe is, I believe, pretty great, due to the immenseness of the area we are considering and the fact that life evolved here on our planet. If it evolved here, why couldn't it have evolved elsewhere? The likelihood that we can locate life elsewhere, though, is another matter. SETI, is a privately funded program, utilizing high technology instrumentation to search the cosmos, in order to do just that, to locate other life...and not to just locate other life, but other intelligent life.

It is thought that all civilizations that have the ability to reason would at some point in their existence, come to understand the electromagnetic spectrum and its myriad of uses. Paramount among these uses is telecommunications, i.e. using parts of the electromagnetic spectrum to communicate over distances with radiowaves, microwaves, or the like.

If a distant life form should happen on this discovery and begin using it then there would be an electromagnetic "signature" that would emanate from the particular planet or area in a particular star group where the planet resides. Instead of just random background radiation that is normally found in space, there would be regularity or sequences of electromagnetic radiation that showed order. This is what SETI does. It combs the cosmos for this type of thing.

Unfortunately, nothing has been gleaned from our efforts yet but the incredible size of the universe and the way electromagnetic radiation is dissipated in space means that we have our work cut out for us. For example, if there was intelligent life in a certain star group that is 1500 light years away, then 1500 years ago, corresponding to our earth's 505 A.D., this distant civilization had to be proficient enough with telecommunications to create "noise" around their planet. Then this "noise", in the form of electromagnetic waves, would need to have sufficient energy to make it as far as our planet and still have the energy encompassed to be detected. We on the other hand would have to have our electromagnetic receiver dish focused on the particular area in the sky where the noise is coming from, to have a possibility of detecting it.

Then, there's the dissipation problem. Just as with a stone thrown into a lake causes rippling and wave propagation in the water that is spreading concentrically from the initiation point out in all directions so too is the energy, from this other planet, moving concentrically away. There is only so much energy in these signals so as the distance from origin increases the energy decreases and hence the likelihood of detection decreases. We would have to have one of our receiver dishes focused perfectly on that star where the planet resides to catch enough electromagnetic radiation to allow the deduction that there is intelligence behind it. Factor in, again, the incredible size of the cosmos, and it truly becomes much worse than just finding a needle in a haystack.

Really, what is hoped for, in order to make things easier on us, is a distant civilization that, like us, seeks to find other intelligent life and is engaged actively in the search. Once again though, due to the great size of the universe, if they are a million light years away from us and started looking at the same time as we started, we won't hear each other for another million years...and both civilizations most likely will be gone.

Herein lay a possible benefit. The odds that some civilization developed, rose, and fell eons before ours makes possible their detection now. Our best hope is to just find evidence from a distant star, regardless of the distance, to show us that at one time in the past, there was someone else out there.

If we did end up trashing our ecosphere to the point that we had to make a Superman-like departure from earth, what would the likelihood be of just finding a new planet to settle? Our galaxy, the Milky Way, contains more than 200 billion stars and an unknown number of planets associated with these stars. It is just one of billions of other galaxies in the universe. Obviously, if we left our planet looking for a suitable planet to inhabit, due to our lifespan, we would be forced to look pretty close to earth.

The nearest star or star group to our solar system is the *Alpha Centauri* star group, of the constellation *Centaurus* visible to us in the Southern Hemisphere. It is made up of three stars: *Alpha Centauri A, Beta Centauri* (also referred to as *Alpha Centauri B*) and the smallest and closest to us, *Proxima Centauri*. These stars are similar to our star's size, the A star being slightly larger and the B star being slightly smaller. They are only 4.36 light years away from us.

I suppose if advances in space travel are made that increase dramatically velocities at our disposal, then we may be able to reach this star group in our life spans. When we got there though, things wouldn't be the same as on our planet since the A and B stars actually rotate around each other and are only 2.2 billion miles from each other (about the distance from our sun to Uranus). The Proxima star rotates around the other two stars, it is thought, in a huge orbit that takes millions of years to complete. Our sun, which is on the small side as far as stars go, is about seven times larger than Proxima Centauri, which barely qualifies as a star. Three stars that close together might make it tough to find a suitable planet but it is thought that several possibilities exist. The Hubble Space telescope is soon to be focused on this group to attempt to find just such a thing.

So, obviously it's rather silly to contemplate leaving the earth due to the vastness of space and the uncertainty of what would be found. It certainly makes much more sense to electively investigate the cosmos rather than having it as a requirement to survival.

Epilogue

The concept of *Deep Ecology* is an offshoot of the ecology movement. The actual phrase was first coined by Arne Naess, a Norwegian philosopher, in 1973. It seems that, according to the basic tenet of deep ecology, the earth has a limited carrying capacity of human population…no more than two billion people, living at feasible levels of technology, can be tolerated. Populations above this cause seemingly irreversible deterioration in the Earth's ability to renew and sustain itself.

Fritjof Capra, Ph.D., an Austria born, American physicist, was on the faculty at Schumacher College, an international center for ecological studies, in England. He used to conduct seminars for corporate executives on aligning businesses more with environmental concerns. I don't know if he's still there but he once stated that "there are hidden connections between everything". He was and probably still is doing his part in educating some of the hierarchical overseers causing the deterioration of the ecosphere.

When discussing Deep Ecology, terms like *ecotage* and *monkeywrenching* come up. As with any ideology there is a radical aspect that insists on being excessively proactive. Ecotage is sabotage motivated by a desire or need to protect ecological integrity. It was a term coined to counter the use of the term *ecoterrorism. Ecodefense* and monkeywrenching are used synonymously with ecotage. All activity in this regard, nearly invariably occurs against corporations thought to be operating without a social or eco-friendly conscience.

Is it corporate America's responsibility, though, to have one eye on the environment and the other eye on the bottom line? Corporations in Western societies prior to the 19th century were merely extensions of Royal or State prerogative, not private economic interests. They were set up to encourage essential activities like the colonization of new territories, the building of transportation infrastructure, the establishment of water supplies, and to provide insurance. These corporations were uncommon and closely regulated.

After America's split from England, corporations at first were limited as they were before but it wasn't long before states allowed a broadening of the scope of self-incorporation throughout the economy <u>for the sole purpose of making money</u>.

At the turn of the 20ᵗʰ century, corporations were thought to exist solely to make their shareholders money. In fact, a landmark case, *Dodge v. Ford*, (Superior Court of the State of Michigan) addressed this very point in 1919. The Dodge in this case was not the automobile manufacturer but minority stockholding brothers, Horace and John Dodge, with a combined 10% stake in the Ford Motor Company. Henry Ford controlled 58% of shares.

In 1916, Ford decided to withhold special dividends paid out to stockholders in previous years in order to pursue philanthropic goals. The brothers took him to court and won. In the decision it was stated that "a corporation exists to benefit its stockholders and that corporate directors have discretion only in the means to achieve that goal. It may not use profits for other purposes".

The message sent to corporate America there was simple, do everything within the law to extract as much money as possible from your venture, for your stockholders. It's no wonder why hard-rock mining ventures and other industries cause and have caused such ecological problems in the past. Looking at the law, strictly speaking, they couldn't, they rationalized, be concerned with the environment because it would cost money which should be going to shareholders. It's the law that should be changed in order to better define parameters for operation.

Things have changed now, though. Court cases have been heard and it has been decided that "expenditures designed to attract customers and advance corporate interests by obtaining 'good will and prestige' also pass muster".

From that it only makes sense that activities that sully the reputation of the corporation would then affect, adversely, the bottom line. In other words, there is a societal responsibility of a business to conduct itself in a responsible manner, if only for the purpose of continuing to make money.

The oil-spill of the *Exxon Valdez* was a black eye on that company and severely affected its bottom line thereby bringing about changes in procedure for transportation of oil.

Cyanide heap-leach mining techniques and their grave ecological after affects in Montana, has led to a near disappearance of hard-rock mining from the U.S.

The trend towards corporations looking at sites of corporate activity with responsibility (as a stewardship) for the environment and hence society, is here. It just needs to occur faster.

Advocates of limiting earth's population to no more than 2 billion and those espousing sabotage of dams and factories to quell pollution occupy the radical fringe of my concept of the Deep Ecology movement. Civil disobedience and tree sitting, in order to educate people on the rate of deterioration of the ecosphere, is even a bit radical. Education through awareness and a desire to know, forms the "grass roots" effort of people truly concerned with our future and what it may hold. Philosophy is at its least a system of thought that governs conduct. It's from Greek for "love of wisdom". We will need all the knowledge and wisdom we can get if we are to continue to inhabit our earth successfully over the next 200 years.

The two most important aspects of Deep Ecology are these: First, we as humans on this earth must change our anthropocentric attitude. We feel that we are the most important life form on the planet and are completely unique, or chosen by God. We may be God's children, depending on whatever you believe, but that shouldn't give us license to subjugate the ecosphere to less important status. It's not here for us to trash. Instead we need to adapt an *ecocentric* attitude, to see humankind as "an integral thread in the fabric of life". Second, we must acquire a new attitude of self-realization. We must attempt to discard our egos and hierarchies and learn to identify more with our ecosphere and all it contains.

As Micheal E. Zimmerman, Professor and Chairman of the Philosophy Department at Tulane University, stated, Deep Ecologists "believe in a less dominating and aggressive posture toward earth if we and the planet are to survive."

Restricting population on the planet to no more than 2 billion inhabitants, although a good idea if considered 100 years ago, is rather impossible without implementing drastic measures today. It seems to me that, given the fact that the population exceeds 6 billion, an attainable low growth rate, guaranteeing numbers won't exceed 8 billion ever, would be more palatable. If that number needs to be higher, then world leaders can decide "how many, by when". By increasing the wealth of Third World countries and raising the level of education everywhere, a consensus can be reached and then implemented. The more important aspect, though, is the effect, per person, of the human inhabitants of the earth. This is where education can have a great impact. Increasing awareness and improving and distributing technology that allows a diminished level of per person impact worldwide is the short term answer.

Whatever is decided as far as rate and limit, the time frame is of critical importance. The problem is that the world ecosphere is relatively harmed now. The clock can't be turned back. Species of plant and animal that are lost, are gone forever. Different areas are polluted to different degrees and possess different recovery rates. Equilibriums in the cycles of life are disrupted and like agitating a snow globe, things just need to settle so new found equilibriums can be established. The ability of the earth to recover from the point at which we corral the deterioration, is important. Scientific advances by way of their helpfulness in solving the problem will be paramount. Guaranteeing, by design, a stable population level, and by worldwide collaborative support for scientific advances, the earth can be revitalized for future generations. Science advancement is the only hope unless, of course, world population just <u>continues</u> to increase. Then it will be just a matter of time.

Inspirational civilizations like that of Tikopia, a tiny island of 1.8 square miles, home to over 1,200 for hundreds of years, yield a note of encouragement. Emergency preparedness, diversity in agriculture, great soil management and space management, and making the tough decision to limit population, has made Tikopia a model.

China, exhibiting the courage and common sense to make the tough decision to limit their population, chose the path that was necessary.

The "every life is precious" mantra is rarely chanted when survivors compete for food during a famine or when heaps of bodies pile up in a pandemic. Tough decisions, regarding limitation of growth rates and education of the masses, need to be made. No society wants to use abortion or infanticide to control populations. No country wants to mandate zero population growth and then oblige their citizens, by force, to comply. By educating members of our ecosphere as to the benefits versus pitfalls of a course of action, there can be acceptance.

Increasing the wealth of impoverished countries, while respecting their culture, is a key element in all this. Merely shoving corporate dollars into a seemingly "backward" country, making it into a mini-America, is not the answer. People can see intent and greed in that approach and will rebel. It can't be by way of religious "proselytization". That infers that the culture being converted, needs to be corrected, which will sow seeds of rejection. The religious approach implies anthropocentrism, anyway, and shouldn't be tried. Too many wars and too many deaths have occurred in the name of one god or another through history, already.

Who is to be the voice of the world? How are we, as Americans, to tell the people in Brazil, who are clearing the rainforest by burning it down, to stop it, when all they are trying to do is provide food for their families and insure a future for themselves. It's especially difficult because we Americans, over the last 150 years or so, deforested a huge area from New York to the Mississippi and from the Gulf coast to Canada. Why should we be able to do this yet tell these people so far away they can't? The only way around this problem is through education, as to the effects on all of us, if we continue our ways. An important part of this will be reforestation programs. Some incredible models of efficiency in correcting effects of deforestation are available to us.

Japan is one of those success stories. They went through a period in their history when they used wood as if it were an infinite resource. This was early in what is referred to as the Tokugawa era, from about 1600 to 1867 (ending 14 years after a U.S. fleet under Commodore Perry sailed into a Japanese port and showed them what arrogance and military might was). The Japanese enjoyed quite a period of prosperity during this time and this was reflected in a population that increased dramatically. With it wood consumption soared. Then there was the great fire of *Meireki*, in 1657. Over half the capital at *Edo*, by some estimates, was destroyed. This disaster created quite a demand, in the aftermath, for timber. It made acutely evident that the policy toward both their forest and their population needed reconsideration. Their continued heavy reliance on wood for both construction and use as a fuel source, would doom their island nation to some future disaster, for sure. Especially if they ignored population increases.

They confronted the problem by educating their populace as to the benefits of a lower population (between 1720 and 1830 the population barely changed!). Within ten years of the fire they had in place an elaborate system of woodland management. They made improvements in construction and heating that dramatically lowered their demand for wood. Their approach was to "tread water" and "buy time". Don't let things get worse and do look for real solutions to the problem.

They started a reforestation program. *Nogyo zensho* was the great *silvicultural* treatise of 1697. In it was the direction and information necessary for an entire civilization to begin treating timber as a crop (just with long "planting to harvest" times). Standardization of seed preparation, germination, and early care after planting, was started. Businesses sprang up to facilitate the process. With these positive approaches came a negative one, also. Japan came to rely more heavily on imported resources in order to conserve their own. This is certainly a problem revisited on many developed countries today, the U.S. chief among them. Japan was very successful though in staving off certain disaster.

Granted, the isolation and relative tranquility aided and abetted the transformation. If there had been competition from abroad and the country was threatened with attack, the transformation might have been impossible. Also, the fact that one family governed for so long during that prosperity, more likely than not, aided the entire process. One can imagine how long the ruling family would've lasted had they not been aware enough to foresee the problem and change direction. Large populations that are unhappy tend to get just a bit testy, even if the members are all the same religiously and ethnically. There are many examples of this throughout history. Taking it so far, though, as Tokugawa did, to the point that the culture was abruptly changed to such a degree, must've taken great courage by the ruler.

Strength of the people controlling the reins of government, is a critical component to all aspects of our survival. Right now because of the factious nature of our world society, consensus on anything would be difficult. The problem of a dying ecosphere is not important enough, now, because the symptoms exhibited by the earth are not costing enough lives or money. As the situation deteriorates there will be less and less distractions. The importance of once all-consuming topics will wither when the "tipping point" is reached. How else can people worldwide be united in a common cause. Just like in the movie *Independence Day*, with Will Smith and Jeff Goldblum, when faced with extinction, humankind will band together.

The problem of birth rate control would certainly be of critical importance, too. This most probably would be the toughest problem of all, to solve. The quandary is the cultural and religious differences and what differing societies would think of population-control solutions that run counter to deep seated religious ideologies. Groups with anthropocentric attitudes and "God's in control" beliefs would be the toughest sell. But surely, even these resisters would see the light, so to speak, if they were in a small enough minority (or if things were blatantly obvious, as they may become).

I know there is a large component in the religious right, at present, here in the United States, that firmly believes that no matter what, God will protect, and if we all should perish, its just God's design. Please tell me, though, that even these *Homo sapiens* would act rationally if provoked by nature, in the right way.

Short term and long term the role of science will be great. Complaints of soaring gasoline and other fuel costs are understandable but moving people out of their comfort zones is usually what motivates them to act. When energy sources become more expensive, our capitalistic system will have an opportunity to shine. Nothing motivates an individual or a corporation more than an opportunity to make money. Gasoline, at $5.00 per gallon, is not that far away. The effect it has had on Europe (they've dealt with double our prices for years) is to promote more efficient means of transportation. The number of small cars, bicycles, and scooters in use on the Eurasian continent, is staggering.

Incentives to derive energy by way of the wind, the tides, the sun, decaying biomass, geothermal sources, and more efficient fuel cells and batteries, will stimulate the sciences. Improvements in technology allowing more efficient use of energy when driving machines, optimal insulation to allow higher levels of conservation, and computer aided aids to every aspect of our lives in which energy is concerned, can be expected.

When all is said and done, every aspect of our lives will have been affected in some way. Successful civilizations in the past, as mentioned before, have transformed themselves in response to ecological threat.

The Inca Empire of South America, with tens of millions of subjects, completely changed their ways relatively abruptly to face a crisis involving population, crop yields, and erosion, not to mention deforestation.

Societies in New Guinea and especially Tikopia offer insight into once ecologically stressed civilizations that have succeeded.

There must have been some "tipping point" in all the successful examples discussed where a "radical state of consciousness" or epiphany occurred. What was the stress in each situation that allowed it to happen? What possibly does the future hold for us the longer we pretend that things aren't as bad as thought?

I mentioned earlier in the Hydrogen section, a concept espoused by Paul and Ann Ehrlich of Stanford University, called the "Rivet Hypothesis"…where diversity in life, regarding both plants and animals, is the glue that holds our ecosystem together. When we get to a point where the world environment is irreparably harmed we will just spiral downward without hope. I actually prefer to think of it maybe as another "allegory" more in tune with the physical world.

It seems that many materials and structures in the physical world have what are called *elastic limits* and *proportional limits*. An elastic limit of a structure is the point at which stress causes a permanent deformation of that structure. A proportional limit is when the linear relationship between added stress and strain is lost and deformation becomes unpredictable and imminent. When filling a balloon with water, both theses phenomena can be observed. The balloon, when compared with other balloons going through similar filling, behaves predictably up to a certain point. When the linear relationship between stress and strain is lost then catastrophic rupturing can happen at any time.

The same holds true for our ecosphere. The warning signs, that the linear relationship between the stress we've been putting into the system and the strain accumulating in the system, are everywhere. The system has gone beyond its elastic limit…stopping now would not return us to our original state. Deformation has occurred. We are somewhere between the balloon deforming in a non-linear fashion and it rupturing. Let's hope we react in time, wake up from our nap, and turn the water off!

Humans…you gotta love 'em. Look at all the fuss they've created. If one critically examines some of their nuances, one may question if there is any hope at all. After all they're just a bunch of members of a species of apes with larger than normal brains. Luckily, through natural selection, they've acquired the ability to learn from history by way of language and recorded documents. They've evolved a moral sense and a way to understand the implications of their actions. There's a mentality of autonomy and interchangeability of interests. There is a relative tendency to conform. Oddly, due to their mass-media generated consciousness, though, morality has become associated with rank, in society. They have this weird concept of celebrity worship. They respond to the strangest stimuli and act in groups that can be both inspirational and embarrassing. They are who they have come to be, though. They have only the tools at their disposal, both mental and physical, and the time allotted. Good luck to them all.

About the Author

I was born into a military family, July 30, 1954, the youngest of four children. My father was a pilot for the MAC (military air command) in the Air Force. From my place of birth, Herlong Military depot outside of Susanville, California, we proceeded to move around regularly.

From Sparks, Nevada which is very close to Susanville, California (where Herlong was) we moved to Fairbanks, Alaska (Ladd AFB). I don't remember anything about it but from the photos I've seen of it, it was very cold. We lived there 2 years until my father's duty station transferred to McCord AFB, in Tacoma, Washington. From here after a few years my father was transferred to McGuire AFB, in New Jersey.

From there we proceeded to Wheelus, Air Base, in Tripoli, Libya. We stayed here 3 ½ years. This was quite an experience, especially given the present post 9-11 climate, since we lived off-base and I, as a child, interacted on a daily basis with Arabs that lived around us or traveled through our area. We stayed here from 1962 to 1965 when my father was relocated to Scott AFB, in southern Illinois. There I attended 6th and 7th grade at a Catholic school and played sports and fished and hunted.

My father then retired from the Service. We relocated to McCord, in Tacoma, where we had purchased a house years before and I attended 8th through 12th grades through the Clover Park School system. I swam while in high school and as a member of the swim team I befriended two very motivated young men, both of whom aspired to attend West Point Military Academy. We conspired to extract a decent education out of the public school system and I planned to follow their lead by attending the Air Force Academy with dreams of becoming a pilot like my father. Fate intervened when, as a senior, I took an early test for eligibility and found out I was red-green color blind. I was told that I could still pursue an education there but that flying was out (I guess there are just too many red and green switches in aircraft cockpits).

I decided to go another direction and enrolled in Tacoma Community College. After earning my Associates Degree I moved on to the University of Washington.

My undeclared majors moved from Journalism, to Forestry, to Marine Biology, and finally to Biochemistry. I declared this major as a senior and through an undergraduate research program also participated in investigations of Protein chemistry. After graduating, I continued my work in the laboratory, working for limited pay from my mentor, Phillip Petra, Ph.D. I was part of a team that studied a protein in the blood that seemed to regulate levels of androgens and estrogens. We used tools like Affinity Chromatography, Polyacrylamide Gel Electrophoresis, and Radio-labeling (with tritium, an isotope of hydrogen). My contribution in studying and characterizing this protein (a beta-globulin called SBP for *steroid binding protein*) was significant enough to earn me an acknowledgement in the study regarding the *Macca Nemestrina* monkey and later two co-authorships for the study of the protein in the dog and baboon.

At this time I had just turned 22 years and was trying to decide if I wanted to attend graduate school to pursue a Ph. D., or go to Medical or Dental school. As it turned out I opted for Dental School and applied and was accepted to the University of Washington School of Dentistry.

During my first year, I applied for and received a grant to study *ultrastructural anatomy* under the tutelage of a Harvard graduated Ph.D. Her name was Margaret Byers. We collaborated with another individual, an oral surgeon, Professor Emeritus, Dr. John Gehrig. We studied teeth removed from humans (for orthodontic purposes), by chemically removing the calcium, staining the teeth with tannin or osmium, embedding samples in *Epon*, and studying the samples with an electron microscope. What we were looking for was the association between nerve cells and odontoblasts (cells that live inside teeth). We became the first investigators to show that this association between the two cells, inside the tooth, was one of adhesion instead of something electrical. This was a significant finding and when I presented the research at the state dental convention, in 1979, as a sophomore, I won first place in the Student's competition, winning a trip to the American Academy of Dental Research convention, in Los Angeles. The research was accepted for publication in the *Journal on Pain*, when I was a senior.

I graduated with my Doctorate in Dental Surgery in 1982 and briefly contemplated continuing in research before succumbing to nesting urges which led to marriage and fatherhood. I have been in private practice in the state of Washington for the past 22 years. I was a member of the part-time faculty in the Dental School Department of Prosthodontics, until 1990. I helped raise 5 children with my lovely wife, Joan, and have been a science and history buff all my life.

I compete in the Washington state Masters swimming program logging roughly 25 miles of pool distance per month. I also like to play golf and travel with my wife (and if we have to take the kids…OK). All my children except one are out of high school. My oldest, Jake, is done with college and is a Real Estate agent, Brandon has just graduated from the UW and is planning to be a teacher, Kailie is graduating soon in medical assisting, Luke is attending Western Washington University in the fall with aspirations of producing movies (but presently is our resident computer whiz; thanks for all your work on the web-site), and finally Jordan is a high school senior, student-athlete (baseball), and appears to be a whiz in chemistry. He plans on attending the UW also.

I owe my interest in history and science to the luck I've enjoyed living in such disparate environments during my early life. Going from Alaska to Washington to the east coast to North Africa in a span of six years was intellectually stimulating, though I didn't realize it at the time. Throw into this the history of the countries or places I visited when young, and you got me.

Sources

Garrett, Laurie. *The Coming Plague* (Penguin Group, Penguin Books Ltd., 1995).
An insightful, comprehensive work on microbes and their role in our lives.

Garrett, Laurie, *Betrayal of Trust* (Hyperion, New York, 2000).
Helpful and insightful during the Bacteria section.

Fenster, Julie. *Ether Day* (Harper Collins Publishing Co., Inc. NY, NY, 2001).
Of particular help for the oxygen section as it pertains to the history of discovery of oxygen, nitrous, and ether.

McPhee, John. *Annals of the Former World* (Farrar, Strauss and Giroux, 1998).
A valuable, exhaustive wealth of geologic knowledge. Especially instructive as it pertains to tools used to study the ancient world. A winner of the Pulitzer Prize.

Cox, Allan. *Plate Tectonics: How it Works* (Blackwell Publishers, 1986)
Written by one of the original "founding fathers' of the theory. A classic.

Emsley, John. *Nature's Building Blocks* (Oxford University Press, 2001).
An invaluable source of information regarding the atomic nature of our lives.

Diamond, Jared. *Guns, Germs, and Steel, The Fates of Human Societies* (W.W. Norton and Company, Inc., 1999).
An incredible look at the rise of civilizations and their interworkings. Also, extremely instructive on the archeological tools used to study the past. A Pulitzer Prize winner.

Diamond, Jared. *Collapse, How Societies Choose to fail or Succeed* (Penguin Group, Penguin Books Ltd., 2005).
An amazing wealth of information about success and failure among historical civilizations. Also very instructive on methods used to study the past and for information regarding Alien species introduction.

Gladwell, Malcolm. *The Tipping Point* (Little, Brown and Company, 2002).
Insight into the way humans think and therefore act.

Medical Microbiology, A Guide to Microbial Infections. 16th Edition. Edited by
David Greenwood, BSC, PhD, DSC, FRCPath, Richard C. B. Slack, MA, MB,
BChir, FFPHM, MRCPath, DRCOG, and John F. Peutherer, BSC, MB, ChB,
MD, FRCPath, FRCPE. (Churchill Livingstone, An imprint of Elsevier Science,
Ltd., 2002).
A final source on all microbiological aspects of my work.

Wilson, E.O. *Sociobiology* (Belknap Press of Harvard University Press, Cambridge, Massachusetts, and London, England, 1975).
This classic work was important in its contribution regarding both human and
animal behavior.

Keeton, William T., *Biological Science*, 2nd Edition (Norton and Company, Inc.,
1972).
Invaluable for basics in processes that govern our world.

Lehninger, Albert L., *Biochemistry: The Molecular Basis of Cell Structure and Function* 2nd Edition (Worth Publishers Inc., 1975).
A great source of biochemical information.

Dictionary of Scientific and Technical Terms, 5th Edition Sybil P. Parker, Editor in
Chief (McGraw-Hill Inc., 1994.

Website:
http:home.austin.rr.com/mikeready/OxygenBindingProteins2.pdf
A great source of binding kinetics as it regards hemoglobin discussed in the oxygen section.

Website:
http://itc.cti.virginia.edu/~meg3c/classes/tcc313/200Rprojs/lavoisier2/home.html
A source explaining the phlogiston theory as well as other aspects of the oxygen
section.

Website:

http://www.sleephomepage.org/sleepsyllabus/n.html

In conjunction with a lecture I went to (Dr. Earl Sommers, DDS, Sleep Apnea and Mandibular Repositioning Devices) an indispensable aid to the sleep section.

Website:

http://www.bacteriamuseum.org/niches/features/diseasehistory.shtml

A great source for information about bacteria.

Website:

http://www.slic2.wsu.edu:82/hurlbert/micro101/pages/Chap1.html#Ancient_History

Historical microbiological information.

Website:

http://www.med.virginia.edu/hs-library/historical/yelfev/tabcon.html

Information on Yellow Fever especially but also other aspect in microbiological history.

Website:

http://www.june29.com/Tyler/nonfiction/pan2.html

A concise history on the Panama Canal.

Website:

http://www.ucmp.berkeley.edu/anthophyta/anthophyta.html

A useful source on the evolution of Angiosperms as well as other related information.

Website:

http://enwikipedia.org/wiki/Flat_Earth

Great information about beliefs of our world in antiquity.

Website:

http://www.asa3.org/ASA/resources/Wiens.html#page%209

Invaluable information regarding isotopes, half lives, and other aspects about this tool used to study ancient times.

Website:
http://www.museum.state.il.us/exhibits/ice_ages/why_4_cool_periods.html
Information for creation of the geology section regarding ice ages.

Website:
http://www.unm.edu/~oberling/cultdom.htm
Information regarding early Neolithic periods and domestication of plants and animals.

Website:
http://www.emayzine.com/lectures/egyptciv.html
Information about the Egyptian civilization.

Website:
http://www.context.org/ICLIB/IC22/Zimmrman.html
Information on various aspects of deep ecology.

Website:
http://www.geocities.com/Athens/Olympus/2547/kura-2.htm
Information about the early great civilizations and their leaders.

Website:
http://www.activemind.com/Mysterious/Topics/SETI/
Useful for the Alien section.

Website:
http://www.star.le.ac.uk/edu/mway/
Great for information about space, our galaxy, and the universe.

Website:
http://www.space.com/scienceastronomy/alpha_centauri_030317.html
Helpful in writing the Alien section.

Website:
http://www.nps.gov/plants/alien/index.htm
Useful in writing about non-native species introduction.

Website:
http://www.botany.hawaii.edu/faculty/cw_smith/aliens.htm
Especially instructive as it pertains to non-native species introduction into Hawaii.

Website:
http://www.wapms.org/plants/hyacinth.html
Useful when discussing non-native species introduction.

Website:
http://octopus.gma.org/surfing/human/zebra.html
Critical information regarding zebra mussel problems as considered in the Alien section.

Website:
http://www.fdrproject.org/pages/toads
Great site for Cane toad information.

Website:
http://www.state.hi.us/dlnr/Snake.html
Very informative for writing the Alien section.

Website:
http://www.fee.org/vnews.php?nid=4170
From John Locke and the John Locke foundation, from "The Heroic Enterprise: Business and the Common Good". Legal information regarding corporate responsibilities discussed in the epilogue.

Index

A

acetylcholinesterase inhibitors, 41
acid rain, 26
acids, 31
adenosine, 42
ADP (adenosine diphosphate), 27, 42
Agassiz, Louis, 89–90
agriculture, 37, 91, 94
AIDS, 74–75
air
    components of, 36
    oxygen in, 53
alcohol, 21
algae, 25
alien life, 113–115
alien species
    beneficial introduction of, 112–113
    cane toads as, 111
    defined, 106–107
    foxes as, 107–108
    on Guam, 111–112
    kudzu as, 109
    Miconia calvescens as, 109–110
    Mile-a-Minute Weed as, 109
    rabbits as, 107–108
    rubber vine as, 109
    Starlings as, 107
    unintended introduction of, 110–111
    woody aster as, 110
altruism, 11
aluminum, 26

Amazon rain forest, 104–105
amino acids
    and digestion, 43
    and DNA, 36
    and nitrogen fixation, 37–38
    nitrogen in, 36–37
ammonia (NH_3), 31
AMP (adenosine monophosphate), 42
anesthesia, 50–52
animal interaction, 11–12
animals
    as alien species, 106–107
    and antibiotics, 64–65
    circadian rhythm of, 18–19
    domestication of, 94–95
    and fatty acids, 7–8
    introduction into non-native
        environs, ix, 107–108
    and poisonous plants, 110
    unintended introduction of, 110–111
anthropocentrism, viii, 119
antibiotics, 33–34, 63–67
antigenic drift, 77
archeology, 102–104
athletics, 20–22
atmosphere, 36, 37, 53
atomic mass units, 24
ATP (adenosine triphosphate), 42
Australia, alien species in, 107–108,
    109, 111
automobiles, vii

978-0-595-35789-5
0-595-35789-X